Special Report 221

Light Rail Transit

New System Successes at Affordable Prices

PART 3 EXTRACT ONLY

Papers presented at the

National Conference on Light Rail Transit
May 8–11, 1988
San Jose, California

Conducted by the

Transportation Research Board

LEARNING FROM OTHER SYSTEMS

Transportation Research Board
National Research Council
Washington, D.C. 1989

Integrating Light Rail Transit into Development Projects on the
Hudson River Waterfront, 121
*Martin E. Robins, Jerome M. Lutin, Alfred H. Harf,
Clifford A. Ellis, and Viktoras A. Kirkyla*

Development and Implementation of Greater Manchester's
Light Rail Transit, 135
A. P. Young

Elderly and Handicapped Accessibility: The California Ways, 147
Robert E. Kershaw and John A. Boda

Ridership Forecasting Considerations in Comparisons of
Light Rail and Motor Bus Modes, 163
Lyndon Henry

Streetcars for Toronto Committee: A Case Study of Citizen
Advocacy in Transit Planning and Operations, 190
Howard J. Levine

PART 3

NEW LIGHT RAIL TRANSIT SYSTEMS AND LESSONS LEARNED FROM START-UPS /199

Infrastructure Rehabilitation and Technology Sharing in
Bringing LRT to St. Louis, 201
Douglas R. Campion and Oliver W. Wischmeyer, Jr.

Hudson River Waterfront Transitway System, 225
Joseph Martin, S. David Phraner, and John D. Wilkins

Alternative Light Rail Transit Implementation Methods for
Hennepin County, Minnesota, 251
Richard Wolsfeld and Tony Venturato

West Side Manhattan Transit Study, 269
*Gregory P. Benz, Wendy Leventer, Foster Nichols, and
Benjamin D. Porter*

Boston's Light Rail Transit Prepares for the Next Hundred Years, 286
James D. McCarthy

Rail Start-Ups: Having the Right People in the Right Place
at the Right Time, 309
Peter R. Bishop

Lessons Learned from New LRT Start-Ups: The Portland Experience, 317
Richard L. Gerhart

RT Metro: Trials and Tribulations of a Light Rail Start-Up, 330
Cameron Beach

Preparation for "Show Time": The Los Angeles Story, 336
Norman J. Jester

PART 4

SYSTEM DESIGN AND VEHICLE PERFORMANCE /349

At-Grade or Not At-Grade: The Early Traffic Question in
Light Rail Transit Route Planning, 351
Michael Bates and Leo Lee

Preliminary Geometric Design Analysis for Light Rail Transit, 368
*Gary A. Weinstein, Raymond C. Williamson, and
Thomas M. Wintch*

RT Metro: From Sacramento's Community Dream to Operating Reality, 387
John W. Schumann

Design of Light Rail Transit Overhead Contact Systems at
Complex Intersections, 408
Willard D. Weiss and Jean-Luc Dupont

Building Light Rail Transit in Existing Rail Corridors: Panacea or
Nightmare? The Los Angeles Experience, 426
Edward McSpedon

Designing to Fit: The Boston Experience, 442
Ronald J. MacKay

Buffalo's Light Rail Vehicle, 449
Ben J. Antonio, Jr.

Pittsburgh's Light Rail Vehicles: How Well Are They Performing?, 458
Ed Totin and Rick Hannegan

PART 3

New Light Rail Transit Systems and Lessons Learned from Start-Ups

Infrastructure Rehabilitation and Technology Sharing in Bringing LRT to St. Louis

Douglas R. Campion and Oliver W. Wischmeyer, Jr.

Metropolitan St. Louis, after 19 years of planning, is developing a dual-mode, cost-effective public transportation system integrating light rail technology with a vastly improved regional bus network. The light rail transit component, known as Metro Link, is an 18-mi continuous fixed-guideway rail line connecting the St. Louis, Missouri, central business district with the Lambert International Airport and McDonnell-Douglas complex to the northwest and with East St. Louis, Illinois, to the east across the Mississippi River. Complementing Metro Link are shuttle bus operations to major employment centers, and realigned routes that form an extensive feeder bus network in the corridor. The initial rail line will directly connect the principal retail, office, recreational, educational, medical, and transportation activity centers with the densest urban population areas. Existing infrastructure is being used, including right-of-way, structures, and facilities to be acquired from two railroads. Nearly all the railroad property is abandoned, but will be revived for this light rail system. Additionally, street and highway right-of-way and other public lands will be made available for permanent Metro Link easements. The capital expense budget for building Metro Link is $287.7 million, covering design and engineering, construction and procurement, testing and start-up, and project management. As a federally funded project, this capital expense is matched with railroad property and facilities acquired separately by the City of St. Louis and donated to the project with a value in excess of $100 million.

D. R. Campion, Sverdrup Corporation, 801 N. 11th, St. Louis, Mo. 63101. O. W. Wischmeyer, Jr., Bi-State Development Agency, 707 N. First Street, St. Louis, Mo. 63102.

FOR NEARLY THREE DECADES most American cities have relied on the conventional urban bus as the primary form of public transportation. Whether riders are transit dependent or riders by choice, their alternative to the private automobile has been and still is, principally, a bus—a form of public transit that must ply the same congested highways, downtown streets, and intersections as the automobile. In larger cities, where population densities and ridership justified a significantly higher level of public transit service, rail rapid transit and commuter rail have continued to serve as major modes of travel in key corridors. Since the 1960s many cities—for example, San Francisco, Washington, Atlanta, Miami, and Baltimore—have developed new rail rapid transit systems.

St. Louis metropolitan-area decision-makers and planners have searched unceasingly since the late 1960s for an ideal solution to the public transit needs of their region. A chronology of events and activities over 17 years led to preliminary engineering for the current project. As the record reveals, St. Louis had its share of false starts and reconsiderations, and St. Louis officials found themselves caught in the ever-changing federal policy maze.

OVERVIEW OF ORGANIZATIONS INVOLVED

In the St. Louis (Missouri-Illinois) metropolitan area two primary organizations are involved in transit planning and programming: the regional council of governments, East-West Gateway Coordinating Council (EWGCC), and the regional transit operator, Bi-State Development Agency (BSDA). In addition, the City of St. Louis and the County of St. Louis (which are completely separate political jurisdictions) are active participants in all transit-related matters.

EWGCC was formed in 1965 as a metropolitan association of local governments. Its two-state jurisdiction includes the City of St. Louis, four counties (including St. Louis County) in Missouri, the City of East St. Louis (Illinois), and three Illinois counties. EWGCC serves as the metropolitan planning organization for the region. The council's board of directors is composed of 14 chief elected officials from local county and municipal jurisdictions; 6 citizens from the region, appointed by elected officials; and the board chairman of the regional transit operator, BSDA. EWGCC is financed by cash contributions (based on a per capita assessment) from member jurisdictions, state contributions, and federal grants.

BSDA owns and operates the regional mass transit system. It also owns and operates the general aviation Bi-State Parks Airport, operates the Gateway Arch tram system, and serves as the regional coordinator for the Port of Metropolitan St. Louis. BSDA was created in 1949 through a compact between Missouri and Illinois ratified by the U.S. Congress. It was given broad powers to plan, construct, maintain, own, and operate specific public works facilities and services. BSDA serves the City of St. Louis, three counties (including St. Louis County) in Missouri, and three counties in Illinois, an area that covers nearly 3,600 mi^2. BSDA is governed by a 10-member board of commissioners appointed by the governors of Missouri and Illinois (five members by each) to 5-year terms. BSDA has no taxing powers, but is a quasi-public agency authorized to issue tax-free industrial revenue bonds, collect fees, and receive funds from federal, state, and local governments.

In 1963, in an effort to stabilize mass transit service in the St. Louis metropolitan area, BSDA was empowered to take over and consolidate 15 separate transit providers. Subsequently, in 1973, a ½-cent sales tax for transit/transportation purposes was authorized by the Missouri General Assembly in the City of St. Louis and County of St. Louis. The city and county annually appropriate these funds in whole (for the city) or in part (for the county) to support BSDA transit operations. BSDA receives support for transit services in Illinois via a downstate transit tax allocation and ¼-cent sales tax in areas of two counties served; both sources of funds are tied to purchase of service agreements annually.

PROJECT HISTORY

In 1983, funding was approved for an alternatives analysis study for the central/airport corridor, which had been shown to be a prime target for major transit investment since 1971. This new alternatives analysis study encompassed five primary alternatives. In July 1984, culminating the alternatives analysis process, a public hearing was held on the draft Environmental Impact Statement (EIS). After receiving all public comments, the EWGCC board adopted a modified light rail transit (LRT) alternative for implementation.

This preferred alternative included LRT between East St. Louis and the University of Missouri's St. Louis (UMSL) campus, all via abandoned or underutilized railroad right-of-way and facilities. The lines then extended to Lambert International Airport and the McDonnell-Douglas headquarters and manufacturing complex at Berkeley (Missouri) via either mixed traffic operation along an existing collector street (Natural Bridge Road) or an exclusive light rail alignment using the Interstate 70 right-of-way. This preferred alternative included a conceptual set of bus service and realignment provisions to effect a feeder bus system to light rail stations and regionwide bus improvements. The estimated capital cost of the light rail component, including more than 18 mi of line, 24 or 25 stations, and 34 vehicles, was put at $250 million in escalated dollars.

The innovative financing developed for funding the preferred alternative was critical to the project's acceptance. The City of St. Louis explored with affected railroads their willingness to provide right-of-way at zero or minimal capital outlay by local government. Compensation for the railroads would entail a swap of the city-owned, and still very much operating, MacArthur railroad bridge across the Mississippi River, public assumption of maintenance responsibilities for railroad bridges in the alignment, and provision of operating rights for one of the railroads on a portion of the acquired line to allow limited freight switching to continue. With an agreement in principle from the railroads to consummate such a transaction, these potential assets were appraised and determined to have a value, if donated to the project, sufficient to cover the 25 percent local-share matching requirement for UMTA capital grant funds under the new start category of the discretionary capital program (Section 3, Urban Mass Transportation Act of 1964, as amended).

UMTA, meanwhile, was expressing considerable reservations about the local decision to pursue the preferred alternative, light rail, rather than the transportation system management (TSM) alternative. Further, although UMTA had provided guidance on the appraisal of railroad assets value, they were not prepared either to accept the appraisal results or to commit to ruling that such assets were indeed eligible to meet local-share matching requirements. But the project's logic, financial feasibility, and uncanny adaptation and reuse of existing infrastructure had now surfaced unmistakably at the local level and in Congress. Through earmarking, Congress designated $2 million in Section 3 funds for a preliminary engineering effort on light rail. Locally, another $1.5 million was allocated from the region's formula allocation of UMTA Section 9A funds, and local cash was raised to provide the match for both UMTA program monies. An application to UMTA for these grant funds was filed by EWGCC in August 1984.

What ensued thereafter was a fairly typical iterative process of application reviews and comments by UMTA. Evidenced in the application review cycle, however, was continued reluctance by UMTA to accept the local decision to pursue light rail. Fortunately, the new budgeting cycle at the federal level was advancing through Congress simultaneously. In anticipation that the St. Louis light rail project would continue to prove its merits through the preliminary engineering analyses and design, Congress acted to again earmark new-start monies for it. The fiscal year 1985 budget earmarked another $10 million for St. Louis; these funds were to be used to initiate final design and construction.

In February 1985 the EWGCC received approval of its grant application to proceed with preliminary engineering on the locally preferred alternative. UMTA, in approving the grant request, stipulated that St. Louis must also evaluate further the no-action and TSM alternatives at the same level of detail as light rail. The EWGCC also agreed to review its demand forecasting, assuring UMTA that the models would be validated (and recalibrated using 1984 on-board survey data from the transit operator) and entirely new travel projections used for preliminary engineering. A final EIS and the UMTA-required cost-effectiveness analysis would also be prepared. The stage and financing for advancing transit improvements were set.

On July 1, 1985, consultants were hired, an EWGCC light rail project office was established, and the preliminary engineering phase was begun—including the additional alternatives analysis and a third demand forecasting techniques assessment.

Demand forecasting techniques were assessed and found to be satisfactory, models were recalibrated and validated, networks for each alternative—including three subalternative lengths of the preferred LRT alternative—were prepared, and travel projections were made. In response to the direction given by the EWGCC board as a result of the draft EIS public comment, analysis of the alternative alignments to reach the airport and Berkeley concluded in the selection from six options of a route that would use Interstate and airport rights-of-way, avoiding any mixed traffic operations on existing thoroughfares and eliminating one or two passenger stations that optional alignments would have required. The initial design work also determined that a major improvement in the alignment in East St. Louis could be made, eliminating in-street trackage and one proposed passenger station. Initial operational analysis also led to a reduction in light rail vehicle (LRV) fleet requirements from 34 to 31 cars, and major changes in the preferred alternative in the downtown St. Louis portion of the line. Detailed modeling work on patron access and egress, and productions and attractions by traffic analysis zone, revealed little negative impact on ridership but substantial positive impact on travel times, and capital and operating costs from the elimination of two underground passenger stations downtown.

All of the preliminary engineering phase activities were augmented and enhanced by third-party oversight. In addition to locally staffed technical, policy, and design review advisory committees that met at least monthly to critique the work constructively, a peer review group and value engineering workshop were convened. The peer review group, composed of seven transit industry professionals from across North America, met at the end of January 1986 in St. Louis to consolidate and tender their critique after several weeks of individual reviews of technical documents. Similarly, a consultant team was given an independent contract to perform a value engineering assessment. This culminated in a week-long value engineering workshop on-site in April 1986.

After 12 months of analysis and design, the preliminary engineering phase was completed. The refined LRT alternative proved through environmental assessment and cost-effectiveness measurements to be the most feasible and prudent course to follow. Engineering and architectural plans were completed to an aggregate 30 percent of design level, with decisions solidified on station locations, track geometry, vehicle requirements and design, construction and procurement contracts and schedules, financing plan, and other deployment details. The initial system of integrated bus services and routes with LRT was defined, detailed, and costed. The time, the option, and the opportunity to deal effectively with travel needs in one key corridor in the region had arrived.

METRO LINK ROUTE

The St. Louis metropolitan area rail transit system, known as Metro Link, is an initial 18-mi continuous fixed guideway rail line from East St. Louis (Illinois) through the St. Louis (Missouri) central business district to the Lambert International Airport and McDonnell-Douglas complex at Berkeley (Missouri). Complementing Metro Link are shuttle bus operations to major employment centers and a realigned regional bus system. The initial line will directly connect the principal retail, office, recreational, educational, medical, and transportation activity centers (see Figure 1).

Metro Link will make maximum use of existing infrastructure. Adaptive reuse of infrastructure is, through rehabilitation of freight railroad rights-of-way and structures, the backbone of Metro Link's feasibility. Included are the historic 113-year-old Eads Bridge (which spans the Mississippi River), the Washington Avenue-Eighth Street railroad tunnel (which runs from the Eads Bridge under the St. Louis central business district), the historic St. Louis Union Station baggage tunnel, a former rail passenger car repair facility and yard, and nearly 14 mi of continuous railroad trackage and right-of-way. Additionally, street and highway right-of-way and other public lands will be made available for permanent, exclusive Metro Link easements. The initial Metro Link alignment will be on a reserved right-of-way, exclusive except for 16 to 18 low-volume street crossings that will be accommodated using common railroad at-grade crossing protection devices.

Because of the availability of existing railroad, highway, and other public rights-of-way, the Metro Link project requires very little real estate acquisition and associated relocation. Near the airport a total of nine single-family residences, all of them under the airport's principal flight path, will be acquired. Elsewhere, only four business properties, three of them at-grade parking lots, will be acquired.

Table 1 displays the Metro Link alignment type and route miles of right-of-way. The existing railroad rights-of-way are being donated by the City of St. Louis to the project after the city has innovatively acquired ownership from

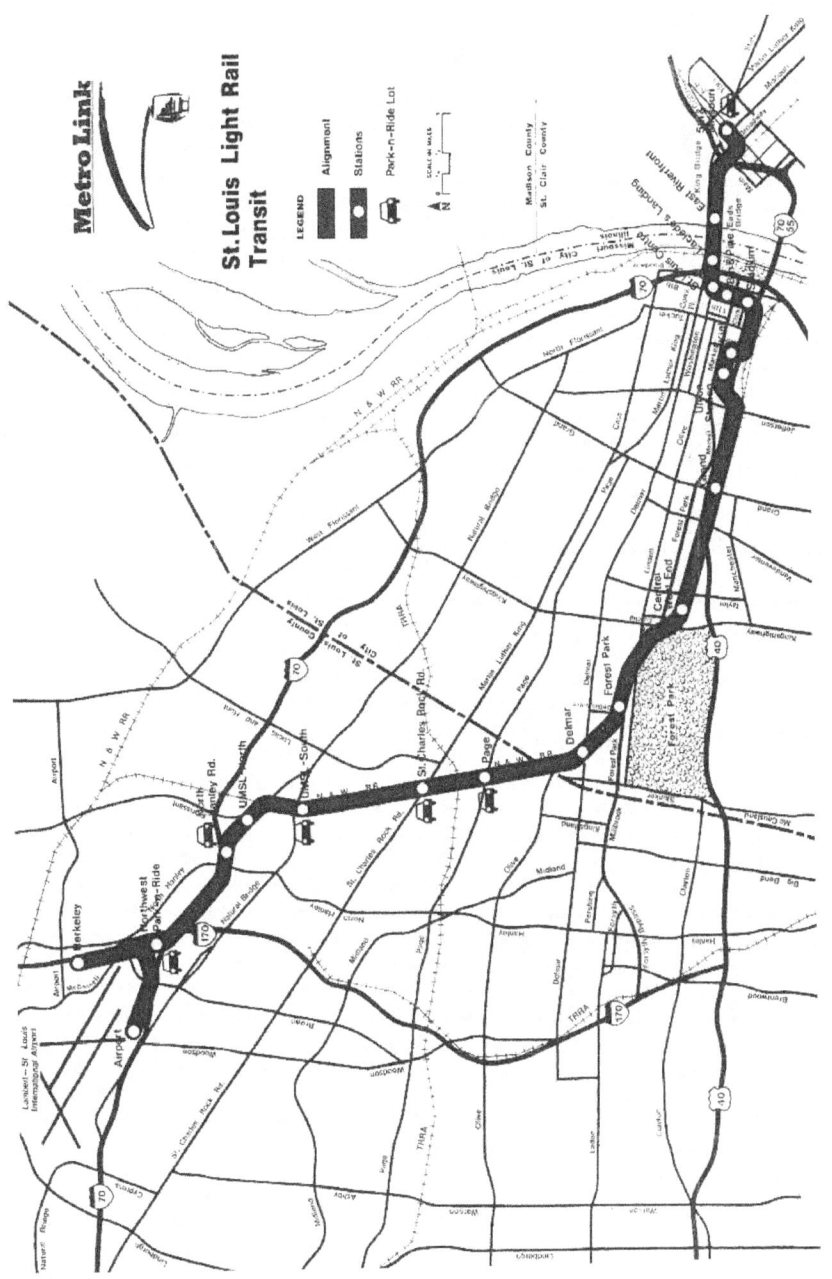

FIGURE 1 Metro Link route map.

TABLE 1 METRO LINK ALIGNMENT/RIGHT-OF-WAY

Alignment Type	Existing R.R. ROW (mi)	Other ROW (mi)	Total (mi)	Percent
At grade	11.8	3.4	15.2	84
Elevated	0.9	0.9	1.8	10
Subway	0.8	0.2	1.0	6
Total	13.5	4.5	18.0	100

the railroads. In the "other right-of-way" category, less than 1 route mile must be acquired from private landowners; the remaining mileage will be made available for exclusive Metro Link use by public entities through permanent, no-cost easements.

DESIGN PHILOSOPHY AND CRITERIA

Not unlike circumstances in the vast majority of its counterpart urban centers across the country, St. Louis had no existing LRT system of its own from which officials could garner practical, local design requirements. The last streetcars in St. Louis ceased operating in May 1966. Consequently, and for better or for worse, the Metro Link managers had to develop a design philosophy without current home-grown experience with light rail. Fortunately, St. Louis came to the preliminary engineering phase with reasonable and pragmatic plans, and at a time when other cities that had already completed a like journey could be tapped for guidance.

At the outset of preliminary engineering the governing charges to staff and consultants were made clear and definitive. Metro Link would be designed based on off-the-shelf equipment, proven technology and construction practices and techniques, strict adherence to budget and schedule, and conscious consideration of every opportunity to incorporate provisions for future system enhancements and extensions. Part and parcel of each of these charges were the overriding goals that the end product be safe, reliable, maintainable, effective, and efficient. Philosophically, then, the initial 18-mi Metro Link system would be capable of being implemented quickly and would provide at least basic rail service that constituents would find immediately successful.

From that rather fundamental and clear project genesis, preliminary engineering proceeded to meet its 1-year completion schedule within its $4.5-million engineering budget and to design a system that, with little risk of overrun, can be deployed for approximately $288 million (escalated dollars) in capital expenditures.

The design philosophy had to be translated into design criteria. To that end, criteria were liberally adopted or adapted from other systems. Because nearly 14 mi of the initial 18-mi alignment are railroad right-of-way with structures built for freight traffic, trackway and trackwork design criteria were fashioned along American Railway Engineering Association (AREA) standards without notable deviation or applicability issues. Systems engineering elements and operational principles were shaped using the San Diego Trolley as a model. Metro Link design criteria for the yard and shops were in large measure an adaptation of Portland's MAX criteria and physical plant.

If there are any elements of the Metro Link design that suggest variation from the U.S. norm for similar projects, they would most likely be station platforms and design implications for contract packaging. After considerable review of what other systems were doing to address the issue of accessibility, and weighing that issue with station dwell times, vehicle dynamics and track geometry, fare collection options, and accident liabilities, Metro Link's designers opted for high-level loading platforms at all stations. Regarding contract packaging, the decision was reached to limit construction and procurement contracts to the smallest number possible—18 contracts at most. Hence, design could proceed in terms of plans, specifications, and estimates in a manner that was conducive to placing the majority of coordinative responsibility on general contractors, not on the Metro Link staff and consultants. Further, the design work carefully disaggregated civil and systems elements so that contract units could be assembled that had the highest likelihood of achieving economies of scale, disadvantaged business enterprise (DBE) goals, optimum equipment, material and labor resource allocations for contractors, etc., within the context of the implementation critical path and right-of-way constraints.

METRO LINK SYSTEM

Stations

Twenty stations will be built along the initial 18-mi Metro Link route. Two will be in East St. Louis, 10 in St. Louis, and 8 in St. Louis County. (The City of St. Louis is a totally separate political jurisdiction from St. Louis County, a century-old circumstance that is not without its negative consequences on fiscal and areawide cohesiveness.)

With the alignment encompassing the reuse and rehabilitation of nearly 14 mi of excellently situated railroad right-of-way, including tunnels and a major bridge, the character of stations was uncontrollable in many respects.

Fifteen stations are at-grade, for the most part accessible without substantive vertical circulation features except for minimal stairs and ramping; one of these stations will be built at the airport to achieve platform interface with the airport terminal's planned people mover system. Three stations are in subway: two in the Washington Avenue–Eighth Street tunnel, and one in the Union Station baggage tunnel. The remaining two stations are on existing elevated bridge structure, one at each approach to the Eads Bridge, where they are enclosed by approach superstructure.

All station platforms are high-level loading to provide full accessibility and to minimize boarding time for all patrons. Platform lengths are typically 200 ft long to accommodate two-car trains. Depending upon the functional and physical location of each station, elevators and escalators will be provided (see Figure 2).

Metro Link stations will be built with materials and finishes chosen with several key criteria in mind. Materials are to be readily available, to have optimal life-cycle costs, and to require only common construction or installation techniques. Station finishes are designed to be resistant to vandalism and to mitigate weathering impacts. Platforms exposed to the elements will have space-frame steel pylon canopy structures with roofing material of copper and glass. Canopies are modular and sized to accommodate 100 percent of each exterior station's peak hour patronage per headway at a minimum of 5 net ft^2 per patron and to cover the complete platform width.

Only essential wall requirements to protect patrons, fare collection equipment, and other elements from crosswinds will be provided, using glass block, free-standing wall segments. The structural elements will be used to support and integrate canopy, lighting, graphics/signage, platform security and communication, and seating requirements. Landscaping will enhance appearance, control and passively direct the movement of patrons within station sites, and enhance or improve microclimates at the stations.

Patron access and egress at stations varies, of course, by location. Six stations will be built with integral park-and-ride lots, providing an initial capacity of nearly 2,000 parking spaces. Kiss-and-ride as well as bus drop-off provisions are incorporated at all station sites except those in downtown St. Louis, where existing thoroughfare provisions adequately perform these functions.

Access and egress treatments are hierarchical. First priority is given to bus patrons using the drop-off lanes, second priority to short- and long-term parking for handicapped patrons and kiss-and-ride patrons, and third priority to long-term commuter parking patrons. Patrons accessing or leaving stations on foot are provided the most direct circulation available to the adjacent land uses.

Page Station

Eighth & Pine Station

FIGURE 2   Metro Link renderings of outdoor and indoor station platforms.

Light Rail Vehicles

As with other federally funded projects, the engineering for Metro Link LRVs has proceeded using a generic car. Conforming to the overall design philosophy, the LRV design used in preliminary engineering was for off-the-shelf, service-proven technology and components.

In this section the generic LRV used in preliminary engineering is generally described. But from this point forward the LRV final engineering will proceed toward completion of a performance specification within a period of 6 to 8 months. That is to say, Metro Link staff and consultants will not design the LRV. Procurement will be based on general and technical conditions that can best ensure proven vehicle and vehicle subsystem performance, leaving detailed design to the manufacturers. Testing at the component level, integrated subsystem level, and, finally, the system level, coupled with pre-revenue and revenue performance criteria, will provide the primary means of product assurance. Also, an on-site maintenance component is planned for inclusion in the procurement to permit the supplier to use his own forces during the first years of revenue service to monitor actual conditions and correct problems that might otherwise cause deficiencies in contracted reliability, availability, maintainability, and other intrinsic threshold levels.

The LRV procurement will use a one-step competitive bid process or, pending further analysis of market conditions, competitive negotiation. In either case, the contract specifications will be aimed at sharing the procurement risks between owner and supplier. Performance criteria, payment provisions, incentives, and damage clauses will be structured to provide owner protection. Supplier control of maintenance for up to 5 years, supplier-detailed design of their off-the-shelf, proven LRV, and the payment and contract incentive provisions will be structured to provide bidder protection.

This procurement philosophy should save scarce resources and time. It will eliminate costly detailed engineering by the owner, whose generic vehicle design constraints under current procurement regulations tend to void much of the work anyway upon bid. Likewise, potential suppliers are given greater latitude in offering a design that they already have and are willing to bid to the performance criteria. They also can avoid costly negotiations over substitutions or equivalents. Being willing to admit that most owners and their engineering consultants are not skilled in manufacturing can pay dividends by reducing final design project costs while simultaneously freeing resources to concentrate on end-product assurance.

This is not to suggest that any and all LRV procurement problems will be avoided, let along mitigated by the Metro Link approach. There are no illusions, only proactive policies that have their roots in the design and procurement experiences of Metro Link project staff and the shared wisdom of colleagues in other transit agencies.

Patronage estimates and the service design require an initial fleet of 31 LRVs. Double-ended, six-axle articulated vehicles with passenger capacity for 64 to 76 seated and 160 to 200 standing at crush load conditions are planned. Dimensionally, the LRV will be between 8 ft 8 in. and 9 ft 3 in. wide, no more than 93 ft long (over couplers) or 12 ft 3 in. high, and equipped with four gangways per side for floor-level boarding.

LRV performance characteristics include maximum operating speed of 55 mph; random and synchronous spin/slide detection and correction control; negotiation of minimum flat horizontal curve radius of 82 ft and minimum vertical (crest or sag) curves of 1,640 ft; and maximum superelevation of 6 in.

Metro Link LRVs will be fully climate controlled, have a normal operating condition interior noise threshold for acceptance of 67 dBA, and general watertightness. Fully automatic, self-centering couplers will be provided for all mechanical, electrical, and pneumatic train connections.

The preceding data are included in preliminary engineering documents distributed in February 1987 to LRV suppliers for an industry review. Very informative and constructive comments were received from every supplier with an LRV currently in service at, or in production for, a U.S. transit agency. These review comments will be revealed at the outset of final engineering. Every performance-oriented criterion or contract condition will be given independent evaluation and reevaluation in the context of both the LRV product requirements and the requirements for interdependent Metro Link project elements. Among other early final engineering tasks, thorough and vigorous integrated value engineering, life-cycle cost, human factors, operations and maintenance cost, and implementation schedule analyses using the largest and longest lead-time contract unit (i.e., the Metro Link LRV) as the catalyst will provide an invaluable project focus.

Yard and Shops

In the planning of yard and shop layouts, thorough consideration was given to all aspects of LRV maintenance, car cleaning operations, operation of the shop with respect to mainline operations, internal operating characteristics, and all other facets of Metro Link-related operating activities. The importance of establishing a clear maintenance and repair philosophy provided the designers with general parameters for a functional, efficient design.

Basic system philosophy consideration and analysis were given to the following requirements to generate specific design solutions:

- Levels of maintenance and repair;
- Work activities;
- Shop loading;
- Contract maintenance;

- Inventory requirements;
- Work flow;
- Space requirements;
- Equipment requirements;
- Personnel requirements;
- Scheduling;
- Routine inspection and preventive maintenance;
- Records, procedures, and method;
- Cost restriction, budget limits;
- Future expansion; and
- Interaction with operations.

An abandoned passenger car maintenance facility and yard on a 10-acre site in the Mill Creek Valley railroad yards area just west of downtown St. Louis, together with two acres from an adjoining city-owned lot, will be Metro Link's yard, shops, and central control location. This site, at the intersection of Scott Avenue and 22nd Street, is approximately one-third of the distance along the initial 18-mi alignment. An existing metal car shed 160 ft long by 67 ft wide by 34 ft high with inspection pit will be rehabilitated and incorporated into the Metro Link shops.

The Metro Link yard and shops facilities will include a three-story maintenance and office building providing approximately 56,500 ft^2 of floor space; a materials storage yard; storage tracks and LRV movement trackage, including a run-around track with a loop; arterial service roads; and parking lots. The yard and shops will handle 24-hr operations.

Three fundamental levels of LRV maintenance, repair, and overhaul will be handled by the shops, i.e., routine maintenance, periodic maintenance, and major repair. Inbound trains from revenue service will be routed to a track or tracks where the following routine maintenance functions will be performed: visual inspections, maintenance technician sign-off, and interior and exterior cleaning. Outbound trains will be inspected by their operators prior to departure. Periodic maintenance will be performed in service and inspection areas, and will include scheduled inspections, correction of deficiencies, scheduled preventive maintenance, and lubrication and testing. Major repair will be done in the shop, including major scheduled maintenance, change-out or complete repair of major LRV components, wheel truing, and collision repair functions. An environmentally separated blowdown facility will be located on a track not normally used for daily inspections.

Space will be provided for the storage of the following types of equipment and structures: electrification poles, signal apparatus, lighting poles, rail, ties, special trackwork, other track materials, ballast, and reels of wire.

Storage tracks initially will provide for 31 LRVs; in the future space will be arranged to accommodate up to 50 LRVs. LRVs will be stored on level tangent track, with both longitudinal and lateral access aisles. Storage tracks will incorporate reused railroad rail salvaged from the existing trackage in the acquired rights-of-way.

Trackwork

The initial Metro Link alignment includes approximately 34 track miles of double-track mainline and one track mile for the airport branch single-track spur. All construction plans and specifications comply with the current edition of the AREA Manual for Railway Engineering and Portfolio of Trackwork Plans, modified as necessary to reflect the physical requirements and operating characteristics of the Metro Link system. Where the system operates across a public street, applicable design requirements of the American Association of State Highway and Transportation Officials (AASHTO), the Missouri Highway and Transportation Department (MHTD), the Missouri Division of Transportation (MDOT), the Illinois Department of Transportation (IDOT), the Illinois Commerce Commission (ICC), and the local counties and municipalities also are utilized.

The track meets or exceeds the minimum requirements of the Federal Railroad Administration (Title 49, Part 213: Track Safety Standards for Class 3 Track). Class 3 track limits freight trains to a maximum operating speed of 40 mph and passenger trains to 60 mph.

The standard gauge of Metro Link is 4 ft $8\frac{1}{2}$ in. Wider gauge will be used in some curves, depending upon the degree of curvature, in accordance with the following: gauge of 4 ft $8\frac{3}{4}$ in. for curves with a degree of curvature greater than 160, but equal to or less than 240, and a gauge of 4 ft 9 in. for curves with a degree of curvature greater than 240.

Primarily ballasted track will be used, meeting the requirements of AREA's Specification for Prepared Stone Ballast. Mainline cross ties will be pressure-treated oak and mixed hardwood 8 ft 6 in. long, conforming to AREA specifications for 7-in. grade ties spaced 20 in. center-to-center on the joint trackage, 24 in. center-to-center in yard track. A ballastless track system will be utilized on the Eads Bridge approach and main river spans and on the floor of the maintenance building at the yard and shops.

All Metro Link mainline track, turnouts, and yard lead tracks will be constructed of continuous welded rail, welded into continuous strings by the electric flash-butt process. Field welds will use the approved preheat thermite process in accordance with AREA specifications.

New rail will be procured for all mainline track, turnouts, and yard lead tracks. Rail will be 115RE section new prime rail, while rail for paved track will be 128RE 7A new prime girder rail. Heat-treated or alloy rails will be used in all special trackwork (i.e., turnouts and crossings) and on all curves where the degree of curvature is greater than 40. The rail for the yard and storage tracks and exclusive freight tracks will be Number 1 relay 115RE rail.

All mainline track with a center line degree of curvature greater than 150 will have an inner restraining rail adjacent to the low rail; rail for this purpose will be Number 1 relay 115RE rail. Emergency guard rails will be installed on tracks on all bridges; for this purpose relay 115RE rail, extending 50 ft beyond each end of the bridge, will be used.

Special trackwork will be manufactured and installed in accordance with AREA specifications and plans. Single crossovers will be used in lieu of double crossovers unless space restrictions dictate otherwise. All special trackwork will be located only on vertical and horizontal tangents; it will not be superelevated. The minimum length between any facing switch points will be 45 ft. The minimum horizontal or vertical tangent distance preceding a point of switch will be 10 ft. Special trackwork is to be located as follows (and includes use of geotextile fabric): Number 10 and Number 8 turnouts with 19-ft 6-in. curved switch points as the standard mainline turnout; Number 6 and Number 4 turnouts with 11-ft straight switch points as the standard yard turnout.

Appropriate measures will be evaluated during the final design of trackwork to minimize stray currents to ground resulting from the use of rails as the negative return for the traction electrification system.

Operations

A track and signal schematic diagram of the mainline route for the St. Louis Metro Link system is shown in Figure 3. The schematic is a simplified representation of station locations, special trackwork junctions, emergency crossovers, pocket tracks, tail tracks, and other operationally important features such as yard locations and railroad junctions.

Trains on the Metro Link system will be operated manually. Signaling and control subsystems are basic and confined to those functions required for safety (i.e., train protection and at-grade street crossing protection) and for the oversight and management of operations at terminals, turnbacks, and transfer zones between yard and mainline areas (i.e., train supervision).

For mainline operations, train protection and supervision are accomplished by these means:

• Train movements will operate by line of sight on Fifth Street in East St. Louis;

• Wayside block signals providing automatic train protection (ATP) will be installed beginning at Fifth and Broadway in East St. Louis and continuing across the Eads Bridge, in the Washington Avenue-Eighth Street tunnel, on

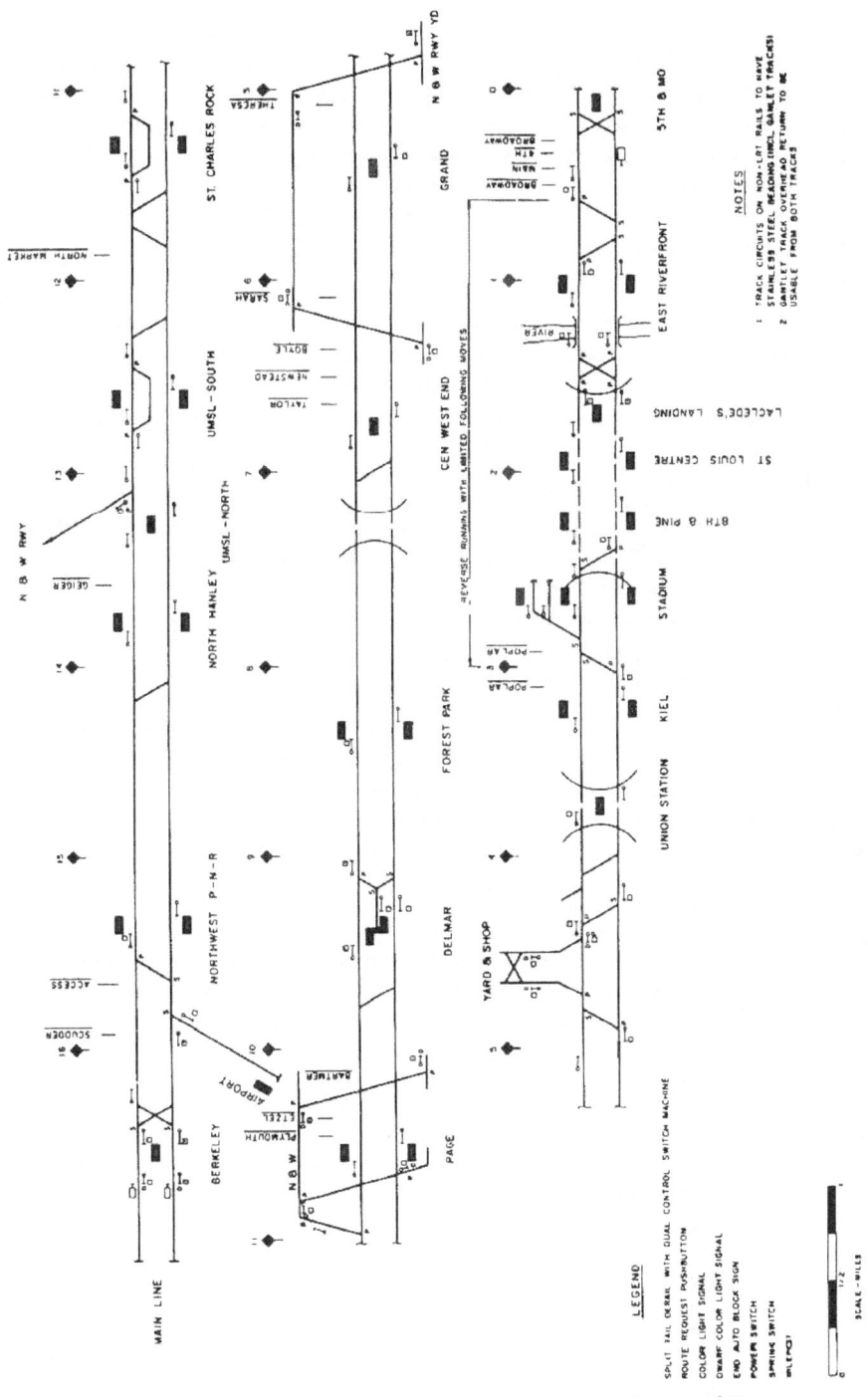

FIGURE 3 Single line diagram of the signal system.

the TRRA/new right-of-way/Norfolk & Western segments from Busch Stadium to UMSL, and on the new right-of-way from UMSL to Berkeley to protect following movements on these high-speed line sections; and

- Signals will be provided on the airport branch to control movements on the single-track section.

Track switches will be controlled in one of three ways. Switches located at junctions where frequent through and diverging facing train movements are made will be power operated, with routes requested by operator-controlled wayside pushbuttons. Switches located in low-speed territory and used primarily for through facing movements and trailing movements from the diverging route will be spring-operated. Infrequently used switches will be thrown by hand.

The 18 street grade crossings along the initial Metro Link line will be protected with railroad-style flashers and gates. Where necessary, crossing protection will be coordinated with adjacent street intersection traffic signals (e.g., at Scudder Road near the airport).

Operations (whether normal or abnormal) will be directed, controlled, and monitored by central control personnel operating out of the shops and office building at Scott Avenue and 22nd Street. Central control will supervise all mainline train operations, maintenance and storage activities, and traction power distribution in accordance with established operating schedules, rules, and procedures. It will implement any corrective actions required to maintain service schedules and to minimize adverse effects of equipment failures or emergency situations. Central control will also monitor station operations to provide for the safety and security of passengers, employees, and system facilities and equipment.

Central control will have several systems at its disposal. The route schematic display system will provide a complete visual indication of the mainline tracks, special trackwork layouts, signal block visual indication limits, and passenger station and substation locations. Radio communications with train operators will permit dispatchers to plot specific train locations manually. The radio communications system will provide channels for train operations, security supervisors, maintenance, and management. Two channels will provide two-way communications between central control and all trains and security personnel. Maintenance and management personnel will have exclusive channels. The telephone system will provide dedicated voice channels for use as telephone extensions from central control to selected sites along the right-of-way, primarily at passenger stations. Telephone service will be provided for passenger assistance and for administrative and maintenance purposes. Emergency telephones will be provided at each passenger station.

The closed-circuit television system will include cameras at selected points in stations and other facilities connected to monitors at central control. The public address (PA) system will be used to issue systemwide announcements (or selective announcements) in all stations. A PA system will also be provided on each LRV so that train operators can make announcements to riders and, via roof-mounted speakers, to people on the wayside. The tape recorder system will provide a record of all dispatcher radio transmissions and phone conversations.

The cable transmission system (CTS) will provide the backbone communication link between central control and various field locations. Terminals located at central control and at each major node of the LRT system will be interconnected by the CTS. The supervisory control and data acquisition (SCADA) system will operate over the CTS. Supervisory alarm and control circuits will connect each fare vendor and each electrical substation with central control. Electrical and support data related to intrusion and field equipment status alarms also will be transmitted on this system.

Trains will reverse direction at Fifth and Missouri in East St. Louis, at the western ends of the line (Berkeley and Airport), and at Delmar and Union Station (21st Street) for turnback service. Train operators will change ends and reset the vehicle destination signs. In addition, at both Delmar and 21st Street, it will be necessary to make diverging moves through the turnback tracks. Turnaround times have been allocated for these tasks.

Speed limits for the Metro Link line are shown in Table 2. These speeds generally reflect performance capabilities, station spacing, adjacent development, and traffic interference. In some locations, sharp radius curves further reduce speeds for relatively short distances.

Normal weekday service (see Figure 4) will begin at 5:30 a.m. and end at 1 a.m. (2 a.m. in East St. Louis to or from Union Station). Commuting peaks will occur from 6 to 9 a.m. and from 3 to 6 p.m.

The number of cars per train is a function of headways, platform lengths, vehicle limits, and street block lengths. The limiting factor for the line is the initial 200-ft platform length, which restricts train lengths to two cars. Two-car consists will be operated on several peak hour, peak direction trains, but

TABLE 2 METRO LINK SPEED LIMITS

Segment	Speed Limit (mph)
East St. Louis to Eads Bridge (East Approach)	20
Eads Bridge (East Approach) to 21st Street	20 to 55
21st Street to UMSL—South	55
UMSL—South to UMSL—North	40
UMSL—North to North Hanley	30
North Hanley to Berkeley	40 to 55
Airport Branch	40 to 55

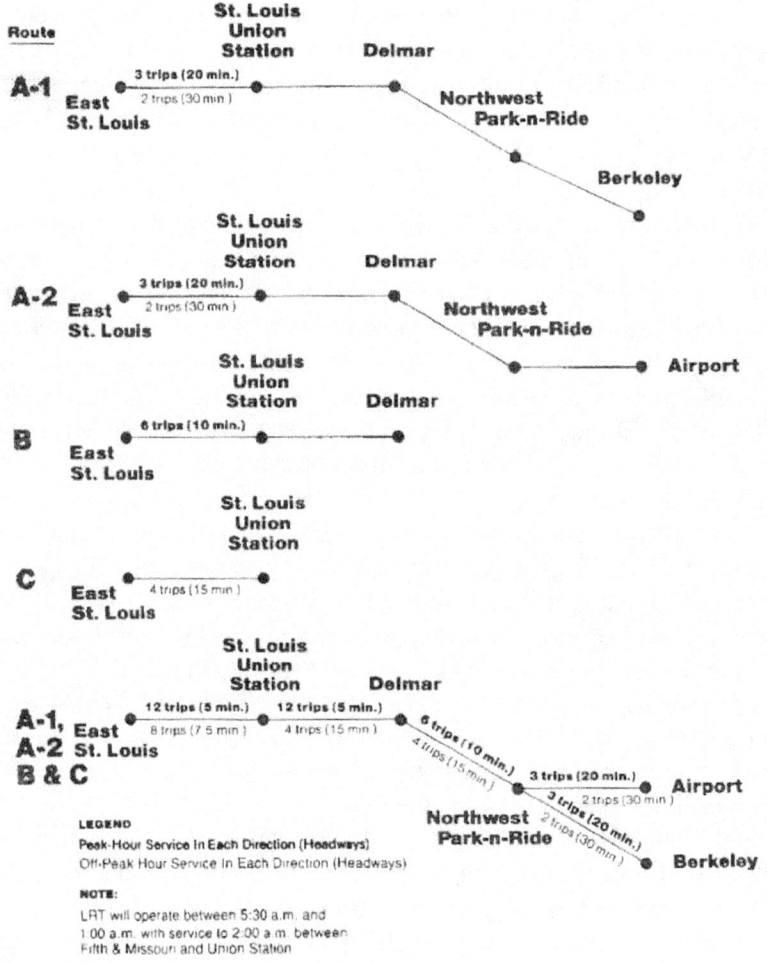

FIGURE 4   Metro Link year 2000 service and headways.

single-car consists will suffice for other peak and all or most off-peak services.

Based on the Metro Link operating plan, including a network of bus routes and services revised to interface with the LRT stations, ridership is projected at about 37,000 daily for the year 2000 (after some seven years of revenue service).

As with other new LRT systems in the United States, Metro Link will utilize a self-service proof-of-payment fare collection system. Fare inspectors will patrol the operation on board vehicles. The San Diego Trolley policy has been proposed in St. Louis as the model legal base for evader citation and enforcement (using the criminal versus civil code).

For system security, a metropolitan transit police force is under review. This police force could work directly for the bus and light rail operator, BSDA, and be augmented by local police departments through interagency agreements.

IMPLEMENTATION AND BUDGET

This section describes the schedule development for the Metro Link project for the final design, bidding, procurement, and construction of all elements of the project. Seven line section construction contracts provide for the basic construction of the 18-mi alignment, the structural elements of the 20 passenger stations, and 6 park-and-ride lots. One station-finish construction contract will provide for the architectural, mechanical, and electrical finish work for the 20 stations. The one yard and shops construction contract will provide for the vehicle maintenance, central control, and storage facility for the system. Four systemwide construction contracts will provide for the trackwork, signals and communications, traction power, and utility relocations. Three procurement contracts will provide for the LRVs, fare vending equipment, and service and maintenance equipment. Other contracts will provide for the consultant assistance for engineering, construction and procurement management, start-up, risk management, and legal counsel.

The schedule gives the sequence for construction and procurement efforts to complete the work, allows 6 months for vehicle and system testing and start-up, and targets revenue service for the end of 1992.

Acquisitions and easements of private properties, railroad properties, and other properties have been or are being finalized early to avoid delaying the construction efforts. Adequate time has been scheduled for long-lead procurements and for the coordination and work of contractors that must complete work within areas of other contracts.

UMTA funding to meet the cash-flow needs of the project to complete work and begin revenue service as scheduled is contractually delineated in a full funding grant agreement, subject only to congressional appropriations under the budget authority contained in the Federal Mass Transportation Act of 1987 (P.L. 100-17).

The final design effort has been organized, and will be completed, in accordance with milestone review and approval dates for 40, 60, 90, and 100 percent submittals for each individual construction contract. Preliminary engineering provided an aggregate 30 percent design level for all work. Therefore, the designated 40 percent review and approval milestone will serve as a midcourse correction checkpoint.

The bidding and award of construction contracts have been timed to provide sufficient time for necessary long-lead procurements and construction activities. The most critical are the design, manufacture, delivery, and acceptance of the LRVs. Other long-lead items have also been considered for their fit into the final design schedule planning. Detailed schedules for the various contracts will be completed early in the final design phase. The anticipated levels of other construction in the St. Louis metropolitan area during Metro Link project construction have been reviewed, revealing no problems in the construction labor market in terms of meeting the project's construction needs.

The systemwide contracts must be completed in partial segments that will coincide with the line segment contracts and their respective schedules, which have staggered starts and time periods. While it will not be possible to start systemwide contracts at one end and progress to the other within the time constraints necessary to meet the anticipated completion date of the project, the general availability of right-of-way will permit these contractors almost unrestrained intermediate scheduling.

The anticipated allocation of funds and the commitment of design and construction dollars based on the contract schedules have been evaluated. The awarded contracts require obligations slightly in advance, on average, of the UMTA grants. However, actual dollars paid out will be well within the UMTA grant cycles each fiscal year. Section 306 of the Federal Mass Transportation Act of 1987 specifically authorizes such advance obligations.

UMTA funds for federal fiscal years 1985, 1986, 1987, and 1988 have been appropriated. The funds for 1989 and beyond are delineated in the full funding grant agreement. This future funding provides a reasonable cushion for cash flow to continue construction to its scheduled completion. Obviously, if the anticipated funds for 1989 to 1991 are significantly varied or delayed, the completion date may be delayed and additional costs may be created for the total project due to continuing inflation additives and other delay costs. Figure 5 shows the capital cost to complete the Metro Link project, $287,699,046. That plus noncash assets contributed at the minimum local-share matching requirement level of 25 percent, or $95,899,682, brings the total to $383,598,728. For comparison purposes, Figure 6 distributes the capital expenditures by common LRT cost elements.

CONCLUSION

St. Louis has attracted nationwide attention by imaginatively recapturing the past and recreating it in modern and exciting fashion. Along the restored riverfront and in the rehabilitated commercial districts and in-town residential neighborhoods, new growth and prosperity have been created by a partnership between public and private interests. A transportation system that sets high standards of quality is needed to continue this revitalization. An LRT system is seen as the cornerstone of this new transportation system.

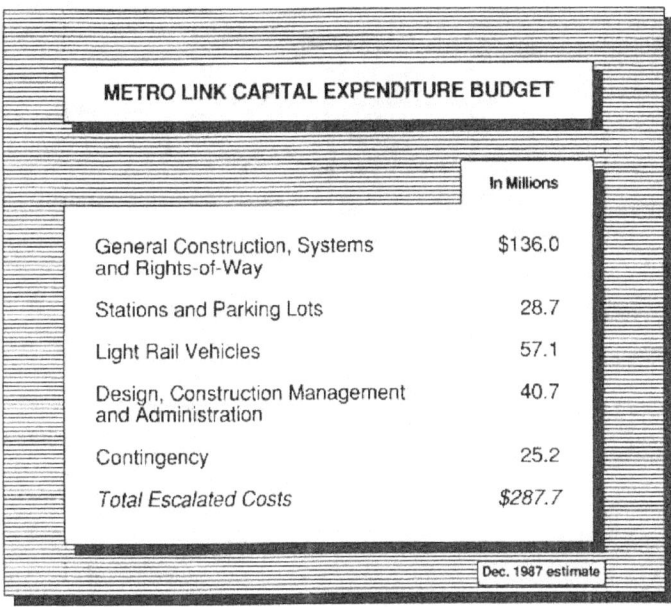

FIGURE 5 Metro Link capital expenditure budget.

In step with cost-conscious times, designers of the LRT system have crafted a practical plan for building this line by maximizing the use of existing bridges, tunnels, and track. This approach on an initial 18-mi line will meet several goals:

- Reduce construction cost by at least two-thirds;
- Virtually eliminate the social, economic, and environmental disruption that typically accompanies large-scale construction;
- Allow for a grade-separated rail operation with higher speeds and fewer delays;
- Reduce or eliminate negative transportation-caused environmental impacts;
- Rehabilitate the historic Eads Bridge and an ideally located downtown tunnel and reuse abandoned and underutilized railroad lines; and
- Ensure an effective core alignment from which prudent extensions can be efficiently deployed to serve every major travel corridor.

The St. Louis LRT project, Metro Link, is on the verge of being built and put into the planned dual mode (bus/LRT), fully integrated mass transit system. Urban rail transit in the region has been a long time in coming back. By simply adopting and adapting proven technical and operational experiences of other LRT systems to the unique alignment opportunity in St. Louis,

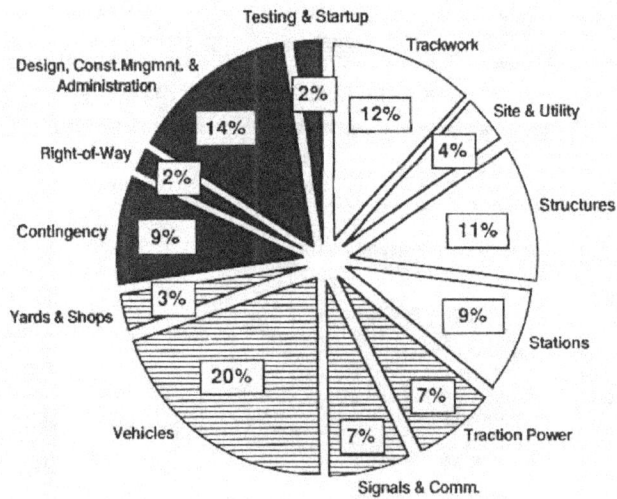

FIGURE 6 Metro Link capital cost breakdown.

Metro Link is feasible and cost-effective. In turn, LRT is the catalyst for a comprehensive restructuring of bus routes that produces a new start for improved public transportation service to the region.

ACKNOWLEDGMENTS

The planning studies for and preliminary engineering of the Metro Link project were conducted under sponsorship of EWGCC in cooperation with BSDA. The resources for undertaking the work on light rail were made available through grants from UMTA and the Missouri Department of Natural Resources/Division of Energy, and appropriations from EWGCC member jurisdictions. The authors would also like to thank the professional staff from the many firms making up the Sverdrup Corporation-led consulting teams, who performed the bulk of the design analyses, and Ellen Towe, whose work was essential in assembling this paper.

Hudson River Waterfront Transitway System

Joseph Martin, S. David Phraner, and
John D. Wilkins

A unique transitway has been proposed for New Jersey's Hudson River waterfront. A narrow strip of land is being converted from railroad yards to large-scale mixed use development. At 35 million ft² of commercial floor space and 35,000 dwellings, this new development requires a high-capacity transitway. Add to the trips generated by the new development nearly 200,000 peak period trips (7 to 10 a.m.) passing through the waterfront to the Manhattan central business district. At least 75,000 trips made by bus ultimately will find their way onto the transitway. The core of the proposed transitway is the state-of-the-art light rail transit (LRT) facility to carry intrawaterfront trips. A busway component and land access roadway have been designated to integrate with the LRT. Transitway design variations include LRT exclusive, busway exclusive, transit in street, bus and LRT sharing right-of-way, and, in one location, bus and LRT sharing travel lanes.

"RECYCLING" IS A POPULAR buzzword in our environmentally aware society. Along the Hudson River waterfront, the term is being applied in two unique ways: recycling waterfront land and recycling the concept of light rail transit (LRT) in support of development. Imagine the opportunities in a strip of land 18 mi long and never more than a mile wide, largely vacant, and 1,000 yd from Manhattan's central business district (CBD). Five years ago, when commercial rentals approached $40/ft² in Manhattan, one perceptive

J. Martin, New Jersey Transit, McCarter Highway and Market Street, P.O. Box 10009, Newark, N.J. 07101. S. D. Phraner, Port Authority of New York & New Jersey, Office of Transportation Planning, 1 World Trade Center, 54N-1, New York, N.Y. 10048. J. D. Wilkins, New Jersey Transit Bus Operations, Inc., 180 Boyden Avenue, Maplewood, N.J. 07040.

developer was purchasing 370 contiguous acres of vacant railroad yard at $21,000/acre less than a mile away in Hudson and Bergen counties, New Jersey, along the Hudson River's west bank.

Development of the Hudson River waterfront renewed interest in LRT in New Jersey. It evolved from a unique combination of changing economic conditions, unusual topography, and dynamic transportation needs. Palisades 150 ft high parallel the river along the northern portion of the waterfront. These cliffs isolate the riverbank from the development on the heights to the west. The narrow strip of land along the base of the palisades is a meager 300 ft wide in some locations.

The first cycle of development commenced in the mid-1800s on reclaimed landfill on the New Jersey side of the Hudson River. Nine railroads established beachheads on the narrow strip of waterfront at the base of the palisades. For these railroads and Public Service Railways (the regional streetcar operator), marine fleets, car floats, and passenger ferries completed the vital trans-Hudson River link. The first development cycle peaked around the 1920s when over 2,000 acres of waterfront were devoted to railroad use. Eight railroad tunnels or cuts penetrated the palisades ridge to serve the waterfront. Public Service streetcars scaled the palisades by various means at eight separate locations. These crossings over, through, and under the palisades were to become strong determinants in sketch-planning LRT transitway alignments.

The first cycle of waterfront development declined when the palisades and river obstacles were overcome by vehicular tunnels and bridges in the 1930s. By the 1960s, waterfront railroad properties lay idle as a result of declining railroad traffic, financial failures, mergers, and abandonments. Five of the largest (and bankrupt) waterfront railroad property owners merged into the Consolidated Rail Corporation (Conrail) in April 1976. Rationalization of Conrail's yards and rights-of-way combined with sale of surplus land by the trustees of bankrupt railroads resulted in hundreds of waterfront acres going on the market. This opened a second cycle filled with land development and transportation opportunities despite the topographical limitations that remained.

Today the challenge facing transportation agencies and land developers is to provide new waterfront transportation overlaid on existing trans-Hudson transportation volumes. Since trans-Hudson services are presently operating at capacity and utilize the same corridors required for waterfront access, staff have concluded that the two markets must be considered together. Officials endorse this dual function concept. A multiagency approach was formed with the New Jersey Department of Transportation (NJDOT), NJ Transit, the Port Authority of New York and New Jersey (PA), and other organizations working together. Partnership with the land developers became a key strategy for bringing transportation capability on line incrementally as development matures.

SOME UNIQUE OPPORTUNITIES

In 1984, a complex sketch-planning process revealed the grand scale of potential development. Even conservative estimates of commercial office space totaled over 30 million ft^2. Waterfront dwelling units at developer-planned build-out would hover near 35,000 units. Analysis confirmed that none of these plans and expectations are achievable absent a strong, visible, high-capacity transit presence.

If developers are to achieve their full build-out plans, the waterfront would have to host 64,000 parking spaces based solely on initial developer expectations. Even with restrained parking policies and high ratios of floor space to parking space (one space or less per 1,000 ft^2), total parking requirements would consume a huge amount of precious space. Nor is there enough roadway capacity to serve anticipated development. Compounding the problem are local land use regulations preserving, among other things, view corridors and view planes from the top of the palisades toward the Manhattan skyline. Placement of towers, size of development, and building height became critical calculations in developer return on investment. Infrastructure either did not exist or was in a state of overload and disrepair. With the exception of Port Authority Trans-Hudson Corporation (PATH), much of the total waterfront area is unserved, even by bus. Rush hour traffic is already congested at levels of service (LOS) D and E because of trans-Hudson and local growth.

The sketch-planning process concluded—and developers recognized—that growth could not be achieved nor could highest and best land uses be realized if automobiles were the primary means of waterfront access. Planning principles devised to guide policy included:

- Suppressed parking;
- Isolation of trans-Hudson and waterfront vehicular traffic flows as far inland as possible;
- Diversion of automobile users to transit in advance of congestion; and
- Trans-Hudson and local bus service and a waterfront transitway system on exclusive rights-of-way.

Fashioning an Alignment

Conrail currently operates its River Line, a freight trunk line, through the Weehawken Tunnel and along the waterfront. This line is of strategic importance to the light rail project because it is a waterfront access tunnel through the palisades and its right-of-way is strategically located at the base of the palisades. The line serves the waterfront from the Weehawken Tunnel south to its crossing of NJ Transit's commuter line into Hoboken. The total length of railroad that can be made available to the transitway system is 4.5 mi, or about 20 percent of the total right-of-way required (see Figures 1 and 2). Fortunately, physical and funding options are available to relocate Conrail to the parallel Northern Branch on the west side of the palisades.

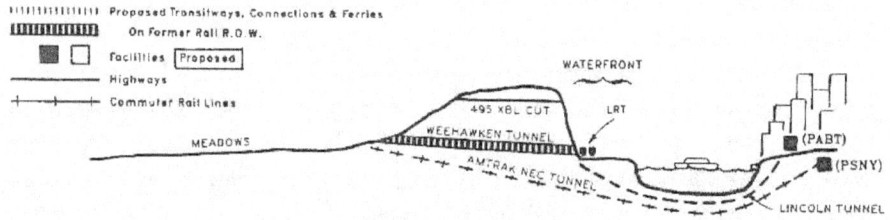

FIGURE 1   Hudson waterfront profile (scale exaggerated).

The state has entered into an agreement with Conrail that will yield benefits that include the relocation and betterment of Conrail's freight operations while vacating the existing River Line right-of-way for its use by the transitway system. The Port Authority's Bank for Regional Development is funding the Northern Branch upgrade and UMTA is funding the purchase of former Conrail waterfront tunnel and railroad alignment. Thus, NJ Transit falls heir to the vacated railroad line for its transitway and NJDOT for its Riverfront Boulevard.

The project has also been fortunate in obtaining a number of easements from private developers who will benefit from the transitway system. Although the construction of the system is some years away, staff approached developers early to ensure that the right-of-way will be available. The first transitway easements were obtained in 1984 from Arcorp. The easement covered nearly a mile of abandoned rail right-of-way north of the Weehawken Tunnel. The agreement was precedent setting, signaling developers' commitment to the transitway concept. Subsequent to that initial acquisition, negotiations with other developers have provided the project with significant amounts of right-of-way in areas where high-density development is taking place. The following rights-of-way have been, or are being, secured without cost to the project:

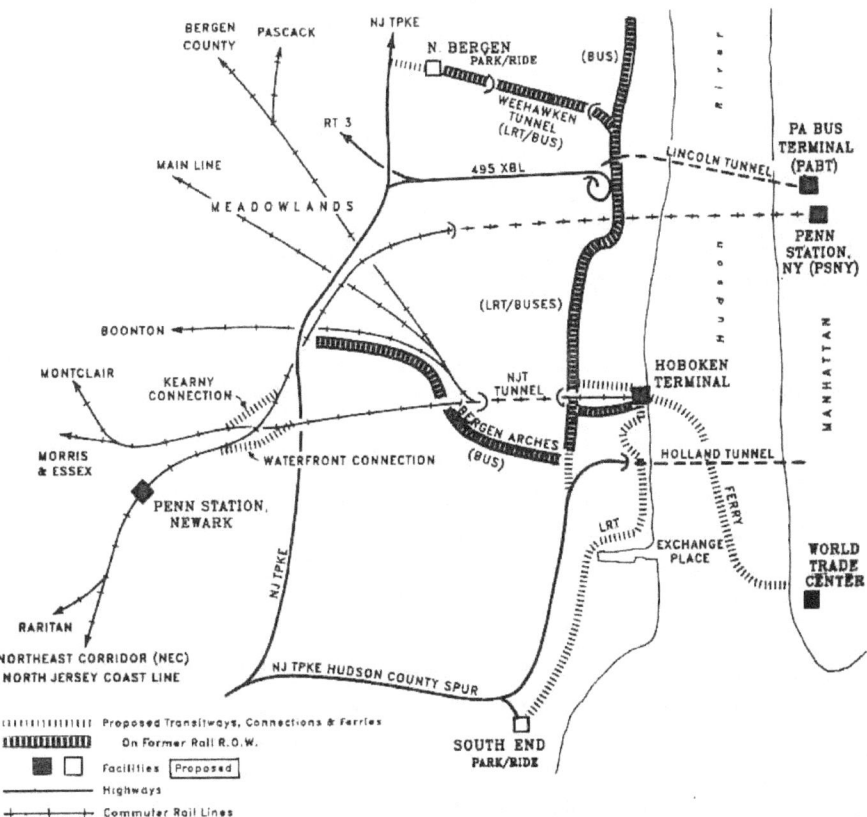

FIGURE 2 Hudson waterfront: existing and proposed transportation.

- Newport Centre—Direct negotiations with this developer yielded a right-of-way across the entire development for a distance of approximately 0.8 mi.
- Lincoln Harbor—Hartz Mountain has provided an additional 30-ft-wide corridor paralleling both its development and the Conrail right-of-way.
- Harborside/Liberty Center/Evertrust—It is anticipated that negotiations with these developers will result in securing a right-of-way in the area immediately north of Exchange Place in Jersey City.
- Lever Brothers Research Center—An agreement has been concluded substituting frontage for former railroad right-of-way as a transit easement.
- Harsimus Cove—Negotiations with this developer anticipate providing rights-of-way to connect the easements furnished by Harborside et al. and Newport Centre.

The combination of the Conrail acquisition with the developer-granted easements is expected to provide the exclusive right-of-way needed for the transitway system where development is densest. Securing transitway easements continues vigorously.

Development Initiative

Development of the Hudson west bank waterfront is on a particularly large scale, although it represents only a modest percentage of the total 775 mi of New York/New Jersey harbor shoreline:

- 18 mi of shoreline,
- 40-plus private and public developers participating,
- 34,900 new dwellings,
- 2,700 acres,
- 32.5 million ft^2 of commercial office space,
- 3.2 million ft^2 of retail commercial space,
- 3,200 hotel rooms, and
- 10-plus marinas.

Heightening the complexity of waterfront development are the institutional involvements. The Jersey waterfront spans two counties and eight separate municipalities, each with its own land use regulations and planning mechanisms. Local jurisdictions successfully defeated attempts to establish a waterfront regional planning institution. To promote development and liaison with developers, the Governor's Policy Office established a Hudson River Waterfront Office. Other state government participants include the Community Affairs, Environmental, and Transportation departments.

Complementary Programs

All transportation programs in the region aggregate to around $14 billion. Several projects are expected to alter dramatically travel patterns feeding to and crossing through the Hudson waterfront. The centerpieces of New Jersey's transportation capital program are two short inland rail connections to unify the two now separate operating segments of NJ Transit's commuter rail system. These connections act like a double slip switch at Kearny Meadows where the Northeast and Morris and Essex commuter rail corridors cross (see Figures 1 and 2). One of these, appropriately called "Waterfront Connection," enables the North Jersey Coast, Northeast Corridor, and Raritan Valley rail services to enter the waterfront directly at Hoboken. Existing and proposed rail services at Hoboken could thereby total 11 distinct rail lines. This, in combination with a Port Authority trans-Hudson ferry proposal, an upgrade of PATH, and the light rail transitway system, creates a waterfront transportation gateway through Hoboken. Prior to the Waterfront Connection, only former Erie-Lackawanna rail services in the northern third of New Jersey accessed Hoboken and the waterfront directly.

One of the major features of the waterfront LRT is the integration of its service with the high-capacity bus, rail, and, eventually, marine modes that surround it. Unlike most other new initiatives, where the LRT is the line-haul service exclusively, this light rail will be designed to perform feeding and distribution for the existing fixed-guideway modes as well as line-haul functions.

WATERFRONT TRANSITWAY SYSTEM CONCEPT

The concept of a joint transitway system that meets the waterfront's transportation needs with LRT and local bus, and trans-Hudson needs with express bus, was based on the planning principles detailed earlier. The transitway experiences in other cities demonstrated a number of options for consideration. Notable is Pittsburgh, where both busways and LRT operate jointly on open right-of-way and through a major tunnel facility. Busways as rapid transit/LRT substitutes in Ottawa service a high-density market, highlighting the capacity and flexibility of this particular mode. Visits to a number of the new LRT properties showed how this mode can be fitted compatibly into various environments.

Existing Highway Transportation

As the map and profile in Figures 1 and 2 indicate, there are exceedingly few access points to the Hudson River waterfront. The mature palisades communities, Hoboken and downtown Jersey City, create effective street barriers of urban density to the west. The principal access routes through the palisades and these communities include I-495, US-1, US-9, and the Hudson County spur of the New Jersey Turnpike. Unfortunately, these access roadways are also the same roadways that are heavily used by vehicles destined to cross the Hudson. These crossing approaches are operating at capacity during the peak hour period.

Local streets through the palisades are other alternatives for reaching waterfront destinations. These streets and boulevards are congested in developed areas. Further local roadway expansion and greater use would only degrade the quality of life in waterfront communities that are in the process of gentrification.

Existing Transit

NJ Transit operates rail commuter services to the Hoboken Terminal from seven rail lines now and may increase that number to 12 in the future. The local bus service operates in a radial fashion from two principal points on the waterfront, Hoboken and Exchange Place in downtown Jersey City. These routes bring riders from locations remote from the immediate waterfront area. With the exception of PATH between Hoboken and Jersey City, these transit services do not now distribute riders along the waterfront. Relying only on PATH raises concerns that it will not have the capacity to service the intrawaterfront market while absorbing more trans-Hudson growth.

Parking

Suburban developers traditionally provide four or five parking spaces per 1,000 ft^2 of office space. These parking ratios are not being incorporated in the waterfront developments. The initial developments along the waterfront have been located near established transportation linkages or planned linkages to New York City. Given this accessibility, the parking ratio at the initial developments has been held down to one or less per 1,000 ft^2 of office space.

Trans-Hudson Perspectives

The trans-Hudson bus system is operating over 700 buses during the peak hour through the contraflow I-495 express bus lane (XBL) and the Lincoln Tunnel to the Port Authority Bus Terminal in New York. This is beyond the practical capacity of the XBL. The ability to provide additional capacity in the I-495 corridor for bus operations is at best temporary. To improve the reliability of trans-Hudson bus service, to reduce total travel time, and to provide capacity for future growth, buses must access their own rights-of-way at some point in advance of the existing congestion. Additionally, the exclusive transitway must be two-way to recycle peak period bus runs, reduce deadhead hours, and handle an expected surge in reverse peak commuting to the new employment generators along the waterfront and other regional attractions.

Functional Requirements

Any waterfront transit plan fulfilling trans-Hudson and waterfront requirements must address four functional roles exemplified by the following trip-end pairs:

- Suburb-waterfront,
- Suburb-Manhattan CBD,
- Waterfront-Manhattan CBD, and
- Waterfront-waterfront.

The lack of good automobile access routes, the inability to make capacity improvements, limited parking, and capacity shortfall of the local street network create a need for fringe park-and-ride facilities. These parking facilities must be located where space, highway access, and direct transit links to the waterfront can be provided. The links to the waterfront also have to perform a distribution function so that persons using the fringe parking facilities have access to virtually all of the developing areas.

Early System Conclusions

A common solution for trans-Hudson problems and the developing waterfront areas was required. These dual needs dictate the nature of transit access to the conceptual Hudson River waterfront transportation system shown in Figure 3. The core right-of-way ingredients that fulfill these combined needs are Conrail's River Line, the associated Weehawken Tunnel, and a back-up penetration of the palisades further south called Bergen Arches (another former rail right-of-way). The Weehawken Tunnel links the waterfront to the Meadowlands, itself a major development area where sufficient land is available for a major park-and-ride facility. Because the Meadowlands area is bisected by both spurs of the New Jersey Turnpike and five state arterials, excellent automobile access will be provided to any park-and-ride facility.

Trans-Hudson bus routes utilizing the New Jersey Turnpike from Passaic, Bergen, and other counties will be afforded easy access to the transitway system by connecting the bus element of the transitway to the New Jersey Turnpike. The specific alignments to accomplish all this are detailed in a following section. A South End park-and-ride is fed off the Hudson County spur of the turnpike. The two park-and-ride lots at the outer extremities of the transitway are expected to provide a viable automobile intercept system. They also feed trips bidirectionally on the transitway.

Initial System Definition

The demand levels and trip concentrations associated with waterfront access needs and intrawaterfront and distribution functions led to the conclusion that a high-capacity LRT would be appropriate for certain portions of the transitway system. This conclusion was reinforced by the high person/trip turnover

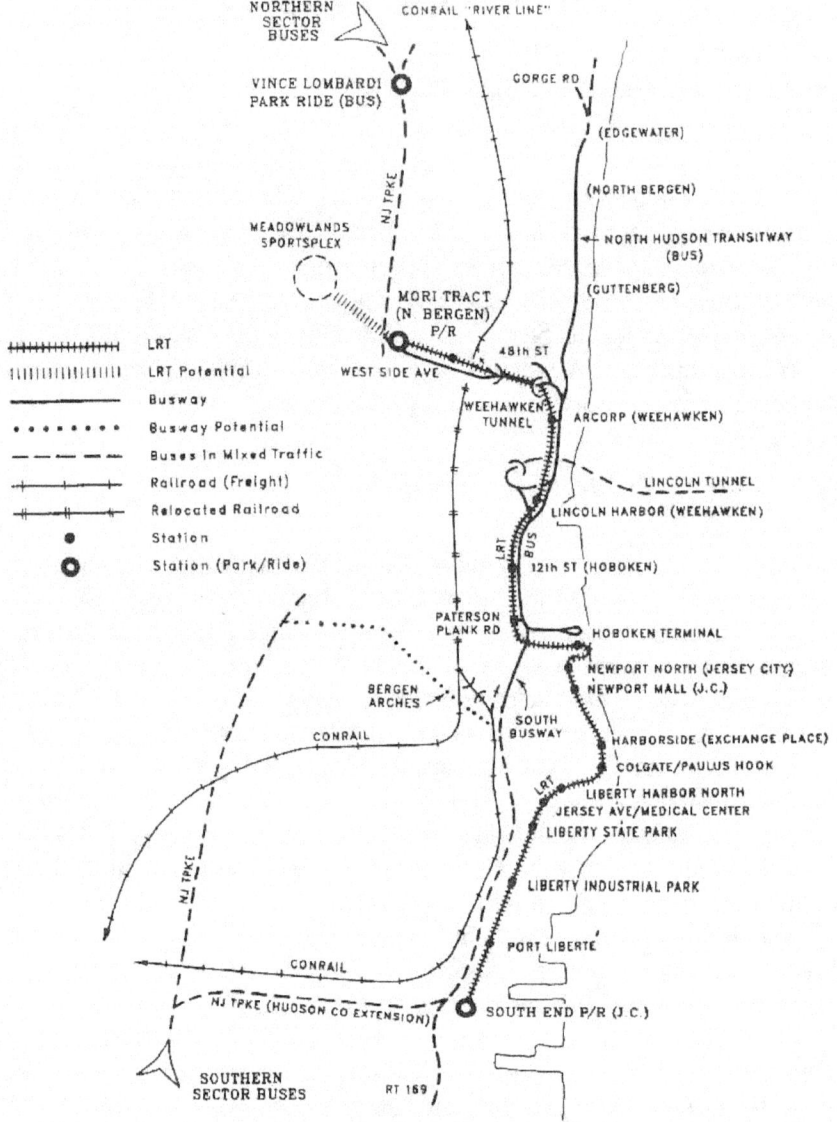

FIGURE 3 Hudson waterfront transitway services and stations.

rate expected at gateway points along a waterfront transit system. Developers were clamoring for a tangible commitment by the public sector to waterfront transportation. They wanted fixed-guideway, permanent, modern, high-capacity transit to complement their "world class" developments—and they appeared willing to help provide for transit that would be uniquely "waterfront."

Next came the determination of which segments would support LRT operations, which would justify busway operations, and which would require joint bus/rail operation. Where joint operations were to take place, staff considered European and North American experiences with various forms of transitways. Pittsburgh's transitway proved the viability of treatments where bus and LRT modes mingle on the same roadway, and where separated modes run parallel within the same right-of-way. But how to adapt joint operation through the Weehawken Tunnel on tight headways proved a challenging traffic management task.

The waterfront system also had to deal simultaneously with express and local service. Both the distributive and waterfront access services are predominantly local-stop in nature. The trans-Hudson services, on the other hand, would stop only at one major interface facility and then operate express to the Lincoln Tunnel portal. This type of operation dictated bypasses for the express trans-Hudson buses skirting station platforms for local transit vehicles.

Based on the vehicles and service types to be blended on the transitway, the following functions and mode pairings were devised:

- LRT Local Services—LRT waterfront services between northern park-and-ride and southern park-and-ride facilities providing local access to the waterfront and an intrawaterfront distributive function en route;
- Busway Express—Trans-Hudson, from northern turnpike connection to Lincoln Tunnel;
- Busway Express—Trans-Hudson, from southern turnpike connection to Lincoln Tunnel (South Transitway);
- Busway Semiexpress—Trans-Hudson, from entrances at Gorge Road and 48th Street to Lincoln Tunnel (North Hudson Transitway);
- Busway Local—from Gorge Road and 48th Street (bus lines servicing northern Hudson County and southern and eastern Bergen County) to Hoboken.

Plotting these functions and modes on a map (Figure 3) reveals a core transitway at the central portion of the waterfront containing joint LRT and bus and joint express and local service. Exclusive bus and exclusive LRT appendages diverge from the core to serve the rest of the waterfront and upland areas.

System Refinement

A conceptual engineering effort further refined a number of issues relating to this project. The major issues included:

- Alignment—What specific alignment should the transitway system follow and what should its specific terminal points be? Where are grade separations required? Are street operations warranted in certain areas?
- Joint Operation—If selected, should LRTs and buses operate in the same pavement area or should they be immediately parallel to one another? What volume and type of joint operation can the Weehawken Tunnel sustain?
- Technology Application—What state-of-the-art bus and LRT technology should be applied to this system?

DESCRIPTION OF THE TRANSITWAY SYSTEM

As presently envisioned, the transitway alignment totals 22 route mi. The total is composed of approximately 13 mi of LRT, 9 mi of busway, and approximately 4.5 mi of joint operation (only in the Weehawken Tunnel do bus and LRT share lanes). This system is depicted in Figure 3. The LRT service will originate at a major Meadowlands park-and-ride facility located on the turnpike either at the existing Vince Lombardi park-and-ride site or at a new site immediately north of Harmon Meadow at what is referred to locally as the Mori Tract. If the former site is chosen, alignment will be oriented north/south, paralleling the New York Susquehanna & Western Railroad. At the south end of Conrail's North Bergen Yard the transitway will turn east to the Weehawken Tunnel. The Mori Tract alignment would originate near the turnpike and proceed east over Westside Avenue to the Weehawken Tunnel. In this instance, provisions would be made for a future extension westward to the Meadowlands Sports Complex about a mile distant.

At the east portal of the tunnel, the alignment would turn south following Conrail's River Line right-of-way along the west side of Hoboken to the Hoboken/Jersey City boundary. At this point, it would turn east to parallel NJ Transit's existing commuter rail line to access Hoboken Terminal.

Leaving Hoboken, the alignment will turn west on an elevated structure for a short distance and then south to serve the Newport, Harsimus Cove, Liberty Center, and Evertrust developments. This will bring the LRT to the Exchange Place area on the surface where access will be afforded to the major Harborside and Colgate developments (12 million commercial ft^2). Continuing south, it will skirt the established Paulus Hook residential area (and historic district) with some street running and provide access to a number of new residential developments along the old Morris Canal basin. South of the west end of the basin, the alignment generally will follow one of several alternative routes parallel to the turnpike to a southern terminus in the Greenville section of Jersey City. En route, the LRT will provide access to a proposed technology center and museum, Liberty State Park, and several residential and industrial areas.

Trans-Hudson buses bound for New York from the northern sector of the commutershed will get new transitway access from the turnpike with an interchange to be built adjacent to the Mori Tract station. Buses would then share the transitway right-of-way with LRT (lanes shared only in the Weehawken Tunnel) to the vicinity of the Lincoln Tunnel. A bus-only link would then be provided for access to the Lincoln Tunnel. In a similar fashion, trans-Hudson buses originating from the southern sector would be diverted initially to the turnpike's Hudson County spur and then operate over the South Busway and shared transitway system to the Lincoln Tunnel. A somewhat longer-range proposal is to build a connection from the turnpike for buses to use the existing Boonton Line and Bergen Arches rights-of-way to connect with the transitway near the Hoboken/Jersey City line.

A busway branch will also be provided along the east palisades north of the Weehawken Tunnel. This North Hudson transitway facility will extend north to Gorge Road and will improve trans-Hudson services for communities in northern Hudson and southeastern Bergen counties. It will also provide a way for closer-in communities to access the waterfront area through the operation of direct local bus service on the transitway to the Hoboken area. The transitway system will provide direct busway access to a new Hoboken bus terminal separate from the LRT. Other local bus routes would utilize portions of the transitway to access the Hoboken Terminal.

System Costs

The conceptual engineering effort nearing completion has generated an estimate of system costs. As described above, the light rail system will cost approximately $638 million; the busway system, $265 million. Table 1 indicates a breakdown of these costs by some of their major components. These costs represent a per-mile cost of approximately $50 million for light

TABLE 1 PROJECT COSTS

Component	Cost ($ thousands)
LRT	410,162
Busway	38,165
LRT/busway	295,249
Roadway	225,948
Right-of-way	89,591
Engineering	154,299
Total	1,213,414

238 LIGHT RAIL TRANSIT: NEW SYSTEM SUCCESSES

rail and $30 million for busway. A review is being made at this time of various design criteria and assumptions that have been made in order to highlight areas where project costs can be reduced.

Ridership

Table 2 indicates the p.m. peak hour ridership for each segment of the line. Maximum peak hour boardings are expected to be 16,379, with 4,163 passengers riding past the maximum load point between the Hoboken Terminal and Paterson Plank Road Station in the northbound direction. The intercept parking facilities accommodate 1,660 riders/hr at the northern facility and 2,847 riders/hr at the southern facility. This table also indicates that one of the prime functions of the LRT is as a distributor, particularly between the Liberty Harbor North station and the Arcorp south station. This table also indicates the major interfaces between the LRT system and the existing bus, PATH, and rail commuter systems.

Table 3 shows the heavy trans-Hudson busway volumes expected on the system in 1995. This level of patronage will compel peak hour bus headway of 9 sec on both the northern and southern approaches to the Lincoln Tunnel. (Present XBL bus headway is less than 5 sec.)

TABLE 2   LRT PASSENGER ESTIMATES: P.M. PEAK HOUR, MORI TRACT PARK-AND-RIDE TERMINAL

	Northbound			Southbound		
	On	Off	Thru	On	Off	Thru
Mori Tract	0	1,660	0	1,065	0	1,065
West Side Avenue	0	1,214	1,660	236	0	1,301
Arcorp	667	1,848	2,874	2,135	663	2,773
Lincoln Harbor	1,052	509	4,055	1,190	327	3,636
12th Street	263	562	3,512	449	277	3,808
Paterson Plank Road	237	589	3,811	255	293	3,770
Hoboken Terminal	2,238	266	4,163	325	1,496	2,599
Newport[a]	968	777	2,191	368	1,060	1,907
Harborside	794	104	2,000	1,315	502	2,720
Colgate/Paulus Hook[b]	864	120	1,310	1,327	656	3,391
Liberty Harbor North	111	24	566	18	234	3,175
Liberty State Park/Jersey Avenue	41	1	479	1	30	3,146
Liberty Industrial Park	40	7	439	3	68	3,081
Port Liberté	78	6	406	5	239	2,847
South End Park and Ride	334	0	334	0	2,847	0
Total	7,687	7,687	N/A	8,692	8,692	N/A

[a]Includes Newport North and Newport Mall.
[b]Includes added trips from Colgate redevelopment.

TABLE 3 PEAK HOUR BUS DEMAND—LINCOLN TUNNEL/XBL

	XBL Approaches			XBL Total	Local Approaches	Total Through Lincoln Tunnel[a]
	Tpke./17	Route 3	Tpke./16E			
1983	235	154	281	670	109	779
1986	266	174	317	757	123	880
1987	272	178	324	774	126	900
1988	278	182	331	791	129	920
1989	284	186	339	809	131	940
1990	290	190	346	826	134	960
1991	296	194	353	843	137	980
1992	302	197	360	859	140	999
1993	308	202	367	877	142	1,019
1994	314	206	374	894	145	1,039
1995	320	209	381	910	149	1,059
2005	368	241	439	1,048	170	1,218

[a]NJ Transit/PA joint venture forecast for bus ridership growth through the Lincoln Tunnel is 36 percent. PA estimate for total trans-Hudson growth from 1995 to 2005 is 10 percent. Anticipated growth for bus ridership is 15 percent owing to the inability of automobile crossing traffic to grow in the same time period.

Stations

As presently planned, there will be 17 or 18 stations on the light rail system. Figure 3 indicates their general locations.

The stations are intended to serve a number of users. Mori Tract and South End stations are primarily intended for park-and-ride patrons and possible transferees. Other stations, such as Arcorp, Lincoln Harbor, Newport North, Newport Mall, and Harborside, are in direct proximity to the residential and commercial developments currently being constructed or planned.

Hoboken Terminal will provide interchange with the commuter rail network, with the Port Authority's planned ferry, and with the existing PATH system. The Hoboken Terminal station hosts bus routes that originate in the palisades communities and can use the transitway. The station at Newport Mall will serve the large 1.5-million ft^2 retail development recently opened. The Harborside station will serve an area in common with PATH's Exchange Place station, a focal point for local bus routes serving the downtown and southern portions of Jersey City. Finally, West Side Avenue, 12th Street, Paulus Hook, Liberty Harbor North, and Port Liberté stations will provide access to both the established and the developing residential and recreation areas along the LRT line.

Three typical station types are being considered, although there will be variations on these schemes to adapt stations to their particular environments.

An LRT station at grade is shown in Figure 4. Platform lengths would initially be 200 ft with expansion capabilities up to 300 ft. Pedestrian crossing would be allowed at controlled points and a station track fence would be installed to prevent random intrusions into the track area. The architectural treatments will support full station accessibility for the disabled.

An elevated station is shown in Figure 5. Dimensions and amenities are similar to the at-grade station. Access to the platform is provided through four stairways located at both the fore and aft portions of the platform. Track fences are placed to discourage intrusions into the track area. The station is fully accessible to the handicapped and includes elevators on each platform.

Figure 6 shows a station designed to handle both bus and LRT vehicles. Light rail vehicles (LRVs) would service joint stations in a manner similar to the LRT-only station except for a merge point between buses and LRVs immediately outside of each station area. Buses would be required to move from the inside lane to access the LRT station platform lane. Express buses would use the inside lanes exclusively and avoid conflicts with LRVs making local stops. Due to the high volumes of buses expected during the peak hours, passenger access to the vehicle lanes is discouraged by design. A center pedestrian barrier stretches the full length of the station to discourage patrons from entering the vehicle lanes. Crossing between platforms will be accomplished by stairways, elevators for the handicapped, and an elevated walkway.

In those areas of the transitway system served solely by buses, station facilities will consist of 10-ft-wide platforms that will vary in length from 80 to 120 ft. Passenger circulation to and between station platforms will utilize at-grade pedestrian crossings as a result of the anticipated lower volume of buses and good sight lines in these areas.

Construction Types

The construction of the waterfront transitway system features several cross-section types to blend it with its environment and to accommodate joint bus/rail operation. In those areas where the LRT operates on its own separate right-of-way, a 50-ft right-of-way will be required as shown in Figure 7. The addition of a busway component requires a total of 60 ft of cross-section (Figure 8). As initially designed, both LRT and busway would share the same roadway in all instances of joint right-of-way use. Based on comments from a peer group review and an in-depth review of the dynamics of accommodating in excess of 400 buses and up to 30 light rail movements in a peak hour, it was decided to separate the bus and rail on a common right-of-way (Figure 9), with one exception.

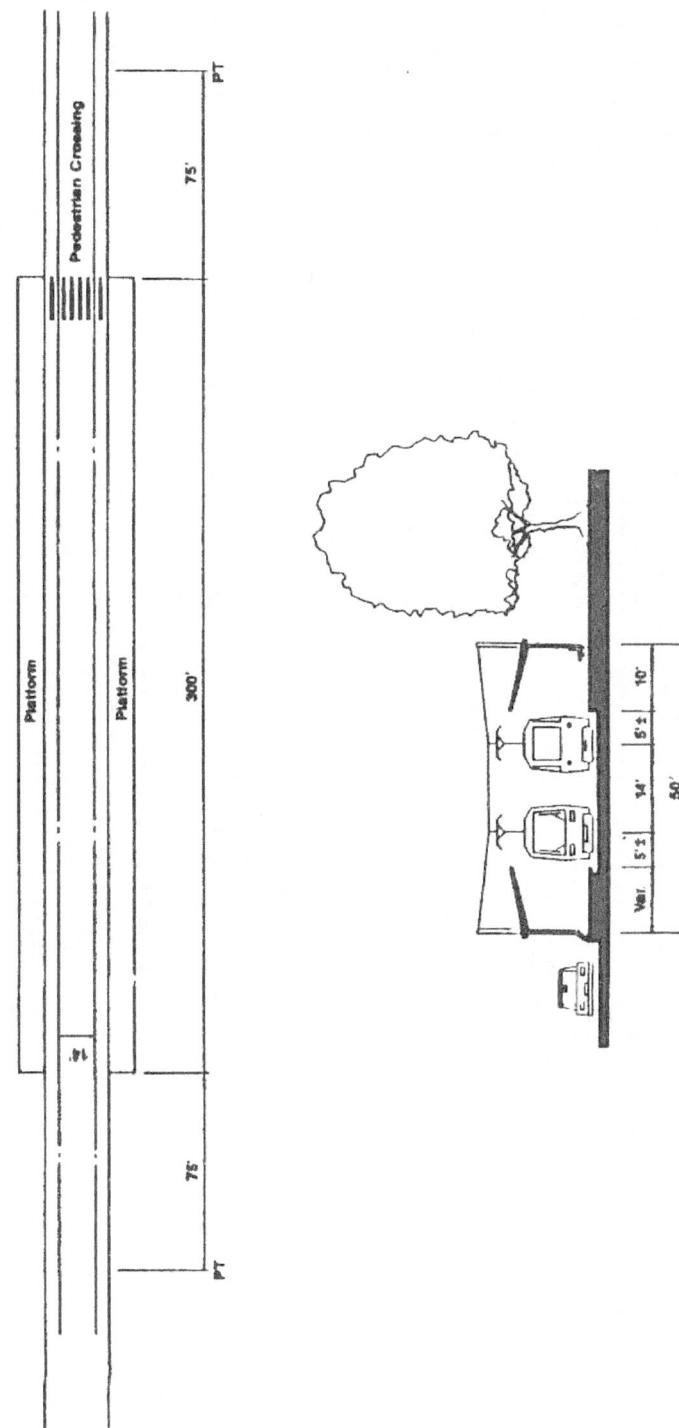

FIGURE 4 Typical station configuration: at grade.

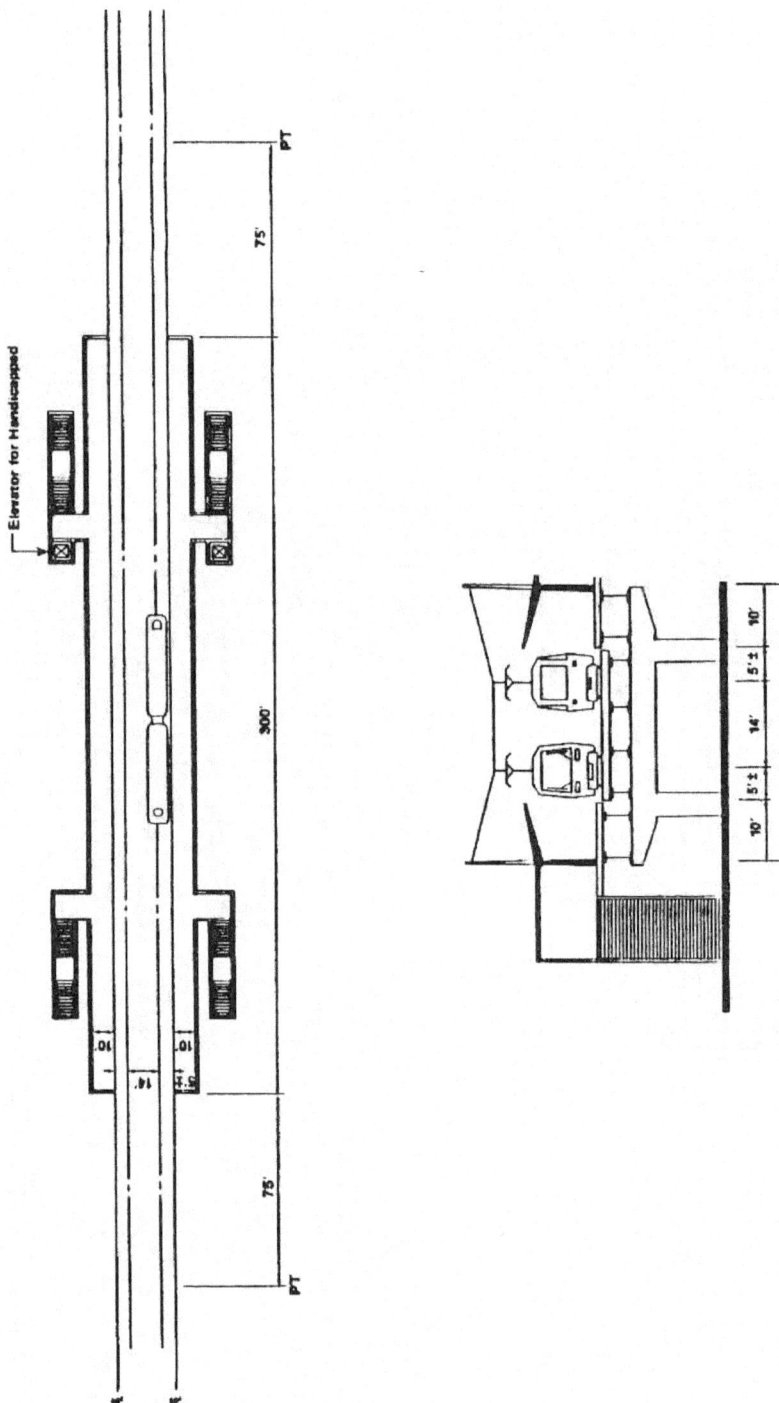

FIGURE 5 Typical station configuration: elevated.

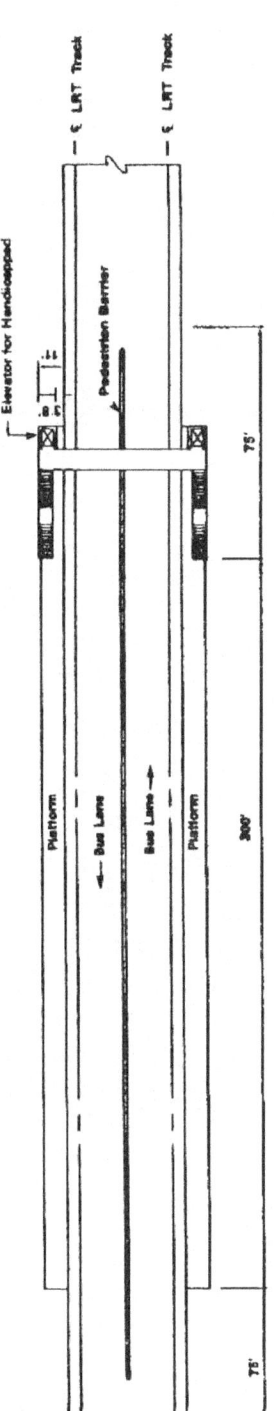
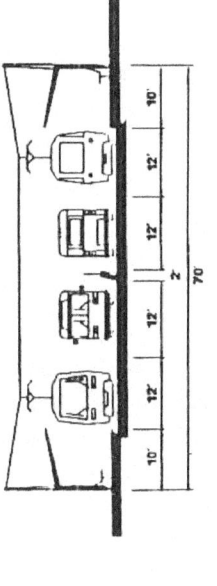

FIGURE 6 Typical station configuration: combined bus and LRT.

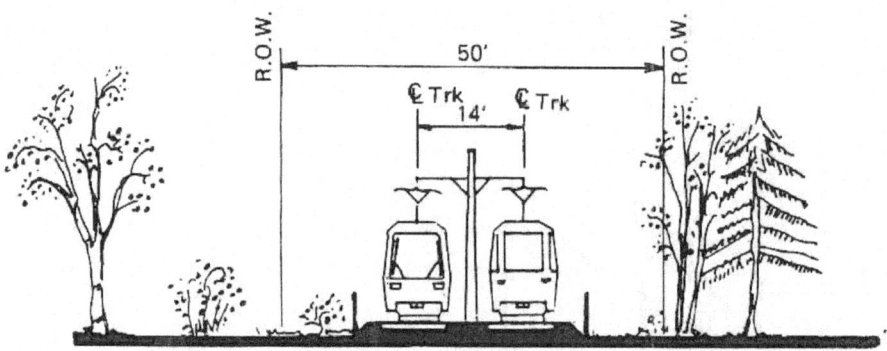

FIGURE 7 Typical cross section: LRT only.

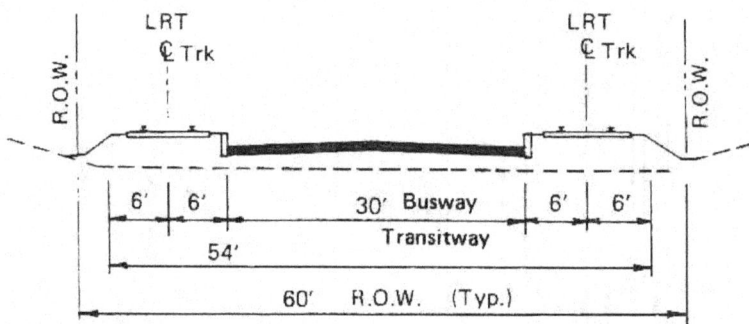

FIGURE 8 Typical cross section: LRT and busway.

Due to the limitation on right-of-way width available through the Weehawken Tunnel, which is only 27 ft wide, LRVs and buses will be mingled on the same roadway through the mile-long tunnel. The cross-section proposed in the Weehawken Tunnel is shown in Figure 10.

In those areas of busway-only operation, the typical cross-section consists of two 12-ft lanes provided together with 8-ft shoulders, and a 10-ft berm. This arrangement is adaptable, though, and could be reduced to 24 ft in areas of limited space.

To the greatest extent possible, the LRT/busway facility will be built at grade to reduce costs. However, there are certain locations along the line where conditions require elevated structures. Elevated locations are as follows:

• From the Mori Tract site to the east side of the Conrail right-of-way—Elevated structure in this area may be the most economical method of crossing the wetlands to avoid a costly earthen fill and accompanying mitigation requirements;

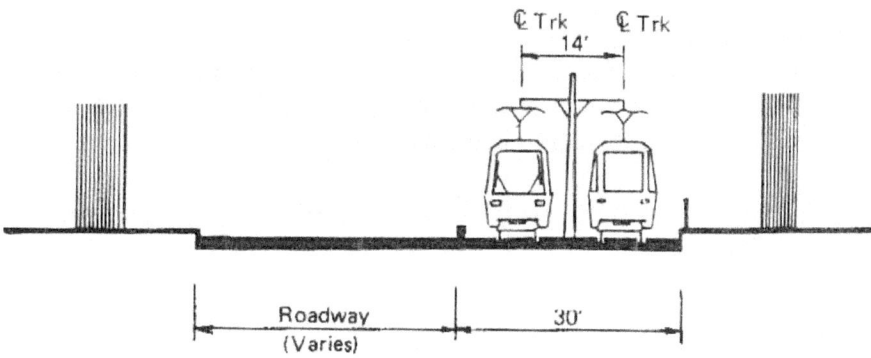

FIGURE 9 Typical cross section: LRT in street right-of-way.

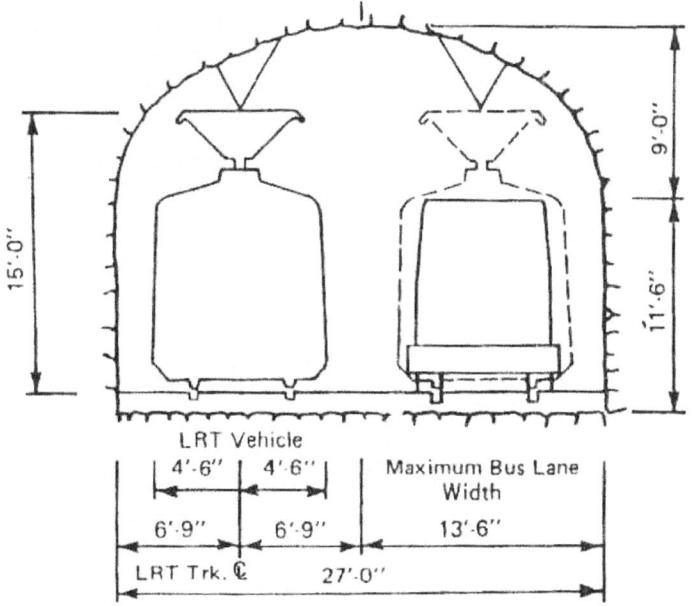

FIGURE 10 Typical cross section: Weehawken Tunnel.

• East of Weehawken Tunnel—An elevated transitway will be provided to grade separate the conflicting merging movements between the transitway routes and the busway from the north;

• Lincoln Tunnel Connector—The busways in this vicinity will be on a set of elevated ramps to sort trans-Hudson bus, local bus, rail, and vehicular movements;

• Crossing Paterson Plank Road and the Morris & Essex Rail Commuter Line;

- Newport—Current traffic projections indicate that grade separations may be required for crossing the major boulevards in the Newport area; and
- Additional elevated structures are being considered between Liberty State Park and the South End park-and-ride facility.

Each of these sections is being reviewed to minimize costs associated with special treatment.

Operating Parameters

Signals and Communications

The LRT system will use a conventional block signal system in those areas where it operates on its own exclusive right-of-way. Traffic signal preemption will be provided as necessary at major intersections. In those areas where both bus and light rail operate on the same roadway or where the light rail is operating within street rights-of-way, line-of-sight procedures will be practiced. The requirement for an on-line communication system will be met by piggybacking the transitway requirement onto the existing state-of-the-art bus radio system.

Transit Vehicles

At the present time the waterfront LRVs are planned to have the following features:

- Six-axle, articulated, double-end units with doors on both sides,
- Capacity for 73 people seated and about 120 standing, and
- 90-ft-long cars with the capability for coupling into two- or three-unit trains with a maximum speed of 45 to 50 mph.

Bus vehicles using the system will include conventional 40-ft transit buses, 60-ft articulated buses in both suburban and city configurations, and MCI commuter buses (intercity design).

Service Standards

During the peak hour, the LRT system will offer initial headways every 3 to 6 min depending upon the consists that are operated. Off-peak headways will be in the range of every 10 to 15 min. The span of service will be approximately between 5 a.m. and 1 a.m. initially, possibly expanded to 24 hours.

Maintenance Facilities

Because of limited available land in the heavily urbanized core of the system, the light rail maintenance center will be located near the northern or southern terminal. Investigations are under way to determine if storage facilities should be split between both ends of the line to minimize the amount of nonrevenue mileage required to set up the daily service pattern. The capabilities of the maintenance facility would be based on those activities already provided by other parts of the NJ Transit system. Integration of the light rail maintenance facility with existing NJ Transit maintenance functions will significantly reduce costs.

Weehawken Tunnel

The tunnel must accommodate both bus and light rail movements. Air circulation will be achieved through the installation of ceiling relay fans to avoid costly ceiling and floor plenums. The design volume for this facility will be approximately 300 buses in the peak hour.

The large peak hour volume of buses through the tunnel, coupled with the difference in braking characteristics between LRVs and buses, requires a unique operating scenario. In the normal operating mode, buses will have free-flow entry into the tunnel. Their bidirectional flow rate will be monitored to prevent more than 22 vehicles occupying the tunnel at any one time. When an LRV is to enter the tunnel, the control system will interrupt bus flow, admit the LRV, and control the time and distance interval between the last bus and any following LRV.

Park-and-Ride Facilities

The terminal park-and-ride facilities are major components of the light rail system. The Mori Tract park-and-ride is being considered in two alternate configurations. The first would feature a five-level parking garage holding 2,860 automobiles. The facility would also enable a transfer between buses and the LRT for those patrons desiring to use the trans-Hudson bus routes in Bergen and Passaic counties to access the waterfront. The conceptual layout of this facility is shown in Figure 11. Another option is to have surface-only parking at a similar capacity. Ordinarily, unstructured parking is cheaper, but the cost of filling wetland areas and mitigation requirements may make surface parking the more costly alternative.

Several options are being considered for park-and-ride facilities adjacent to the southern terminus. In all cases, access would be provided to the Hudson County extension of the New Jersey Turnpike and other arterials.

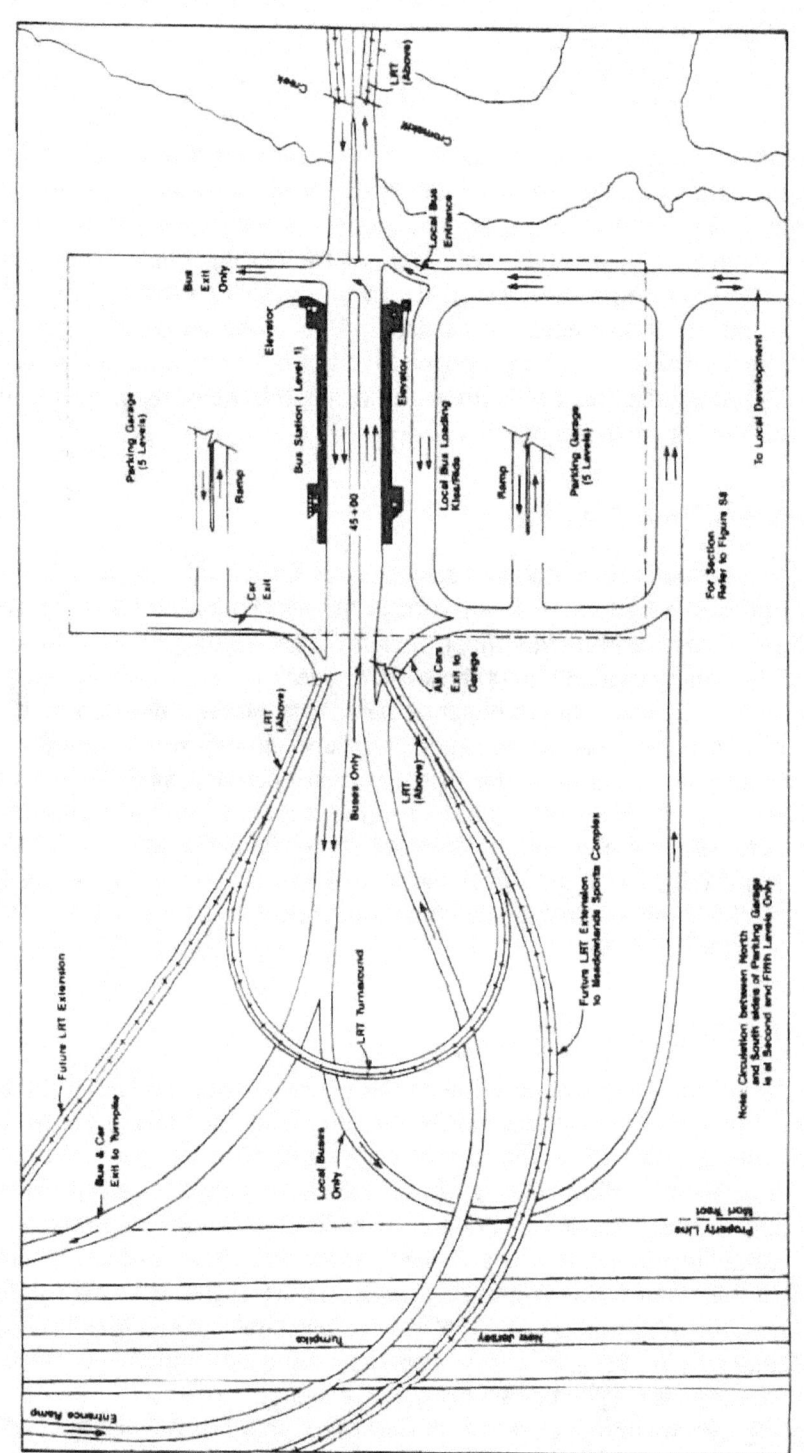

FIGURE 11 Mori Tract park-and-ride.

Funding and Institutional Roles

Funding initiatives and precedents are under way along several fronts. To meet its transportation capital needs, New Jersey has established a Transportation Trust Fund derived from gas tax revenues. This initiative, approved by the legislature in January 1988, is intended to address New Jersey's comprehensive travel needs, including the waterfront. Federal funds have already been applied to right-of-way acquisition along the waterfront. The Port Authority established two dedicated regional development funds from which New Jersey and New York each can draw at their discretion. New Jersey has already withdrawn funds for waterfront transit, highway, and pedestrian walkway projects. Finally, the developers have contributed rights-of-way and, in some cases, agreed to share the costs of transit improvements on the rights-of-way. The following institutions have already contributed to the study and design effort or supported right-of-way acquisition aggressively:

- NJ Transit Waterfront Office—Has been lead agency charged with overall responsibility for planning, design, and acquisition of the transitway system along with financial planning;
- New Jersey Department of Transportation—Provides engineering support for the planning and design effort; negotiates right-of-way acquisition with their consultant, Parsons, Brinckerhoff, Quade and Douglas; sponsors the initial study and design reports;
- Private Developers—Have granted dedicated right-of-way easements and other considerations through their properties and coordinated their designs;
- NJ Transit Bus Operations—Is proposed operator of the transitway property with major role in design standards and bus operations planning;
- Port Authority of New York and New Jersey—Has provided funding assistance for relocating Conrail off the waterfront, initiated consideration of several busway segments in sketch-planning phase, provided technical assistance on bus element of transitway and XBL bypass, and is providing funding assistance on South Busway segment of the transitway;
- Governor's Waterfront Office—Has played major institutional role in advancing the project and liaison with local jurisdictions, resolves land development and transportation issues, and participates in design;
- UMTA—Has provided funding for acquisition of Conrail's waterfront right-of-way to form the transitway core (further federal assistance is anticipated);

- Local Jurisdictions—Have adjusted plans and regulations and provided assistance through waterfront advisory body and directly on local problems; and
- Statewide Authorities and Private Institutions—Have provided other funds.

CONCLUSIONS

The last major addition to the North Jersey rail transit system occurred on May 26, 1935. On that date, Newark's City Subway opened as a light rail operation and closed an era of rail transit expansion. The City Subway, as a concept, an institution, and a light rail property, survived while other rail services in the New York/New Jersey region were discontinued. It is significant that this last new addition in 1935 and the anticipated future addition, the waterfront transitway of the 1990s, are both light rail.

Alternative Light Rail Transit Implementation Methods for Hennepin County, Minnesota

RICHARD WOLSFELD AND TONY VENTURATO

The Comprehensive Light Rail Transit (LRT) System Plan for Hennepin County, Minnesota, defines a Stage 1 system, a 20-year system, a financial plan, and an implementation plan. The purpose of the implementation plan is to define the contractual relationship between the Hennepin County Regional Railroad Authority (HCRRA) and the suppliers of the LRT system, to define the system operating and maintenance responsibility, and to define the relationship between associated land development and the LRT system. The reason for investigating alternative implementation methods is that much interest exists in involving the private sector to the maximum extent, consistent with the public interest. LRT system implementation will include not only the construction and procurement of system facilities and equipment, but also the financing of this work. In addition, options may be available to involve construction and procurement contractors in the operation and maintenance of the system after it is built. Recent years have also seen great interest in coordinating land development with rail transit construction. In some instances, developers of adjacent land have participated in the financing of transit stations. This report defines the LRT system components, identifies and evaluates alternative implementation methods, and outlines conclusions on an approach to LRT system implementation.

THE TWIN CITIES OF Minneapolis and Saint Paul have analyzed the feasibility of fixed-guideway transit systems since 1968 when the Metropolitan Transit Commission was formed. To date, no system has been implemented. The general history includes the following events:

R. Wolsfeld, Bennett, Ringrose, Wolsfeld, Jarvis, Gardner, Inc., 700 Third Street South, Minneapolis, Minn. 55415. T. Venturato, Bechtel Civil, Inc., 3505 Frontage Road, Suite 250, Tampa, Fla. 33617.

- Late 1960s and early 1970s—studies resulted in a regional, preferred fixed-guideway system that was automated and used a 40-passenger vehicle;
- Mid-1970s—significant study of personal rapid transit was undertaken and studies of busways were completed;
- Late 1970s and early 1980s—feasibility studies of LRT were completed;
- Early 1980s—Minnesota legislature enabled counties to establish regional railroad authorities to implement light rail transit (LRT) systems;
- Mid-1980s—implementation studies of LRT were completed; and
- 1985–1987—the legislature prohibited any public agency from spending public monies on the planning or design of LRT.

These studies were all undertaken by a metropolitan or state unit of government. This governmental focus changed in 1987, when the Minnesota Legislature reinstituted the authority of regional railroad authorities to implement LRT. Thus, after many years of discussion, a single agency has the authority to proceed with LRT implementation in Hennepin County. The legislation required that the railroad authority prepare a comprehensive LRT system plan. The plan is based upon previous work and answers the following questions:

- Where will LRT services be provided within 20 years?
- What will the Stage 1 system include?
- What method will be used to implement the LRT system?
- How will the LRT system be financed?
- Who will operate the system?

POTENTIAL LRT ROUTES

Five corridors considered to have high potential for successful LRT implementation are being analyzed in the LRT system plan (see Figure 1). The Northwest Corridor connects downtown Minneapolis with the northwestern suburbs. A 1981 study of LRT feasibility in the region identified this corridor as a high priority. The University Connector would link downtown Minneapolis and the Minneapolis campus of the University of Minnesota, the third largest generator of transit trips in the region. This link was a portion of the corridor including downtown Minneapolis to downtown Saint Paul, which was studied in a recently completed alternatives analysis.

The Hiawatha Corridor would connect the downtown area with the airport and the proposed 10 million ft^2 of development known as the Mall of America in Bloomington. LRT was identified as the preferred transit alternative in a corridor Environmental Impact Statement (EIS) completed in 1985.

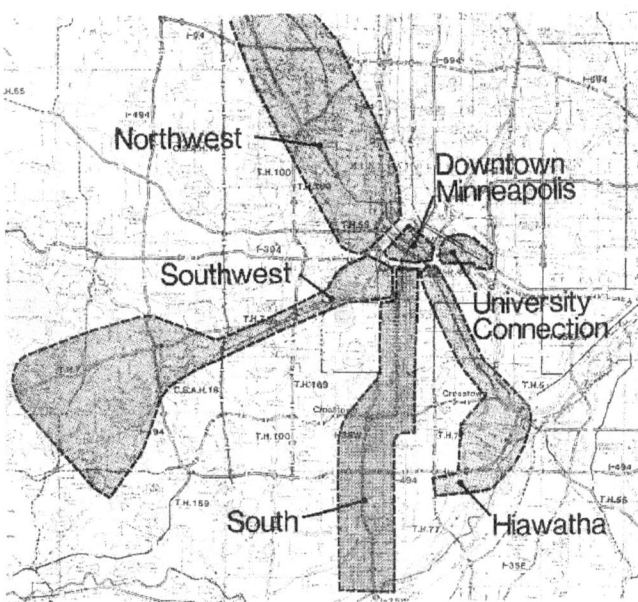

FIGURE 1 Study corridors.

The preferred roadway component of the corridor plan, a four-lane at-grade arterial, will go under construction in summer 1988. The South (I-35W) Corridor would connect south Minneapolis and the southern suburbs of Bloomington, Richfield, and Edina with downtown. Concurrent studies by the Minnesota Department of Transportation (MDOT) and the Metropolitan Council are assessing the need for I-35W highway and transit improvements in the corridor. The southwestern suburbs are connected to downtown via the abandoned Chicago and Northwestern Railroad right-of-way purchased by the Hennepin County Regional Railroad Authority in 1984. LRT was considered as an alternative during a draft and alternatives analysis of the University Avenue and Southwest Corridor.

In June 1988, the Hennepin County Regional Railroad Authority (HCRRA) adopted a Stage 1 system and a 20-year plan that are shown on Figures 2 and 3 and summarized in Table 1. The major conclusions of the plan are to proceed with implementation of an LRT service in the Twin Cities, to provide service in multiple corridors in Stage 1 versus service in a single corridor, and to construct a tunnel in the downtown area.

The financial plan for the Stage 1 system includes a countywide property tax, tax increment around stations, a portion of state sales tax on motor vehicles, and private sources. The county currently levies 0.75 mill, which raises $7 million per year. In April 1988 the Minnesota Legislature appropriated $4.17 million from the sales tax on motor vehicles for LRT "planning,

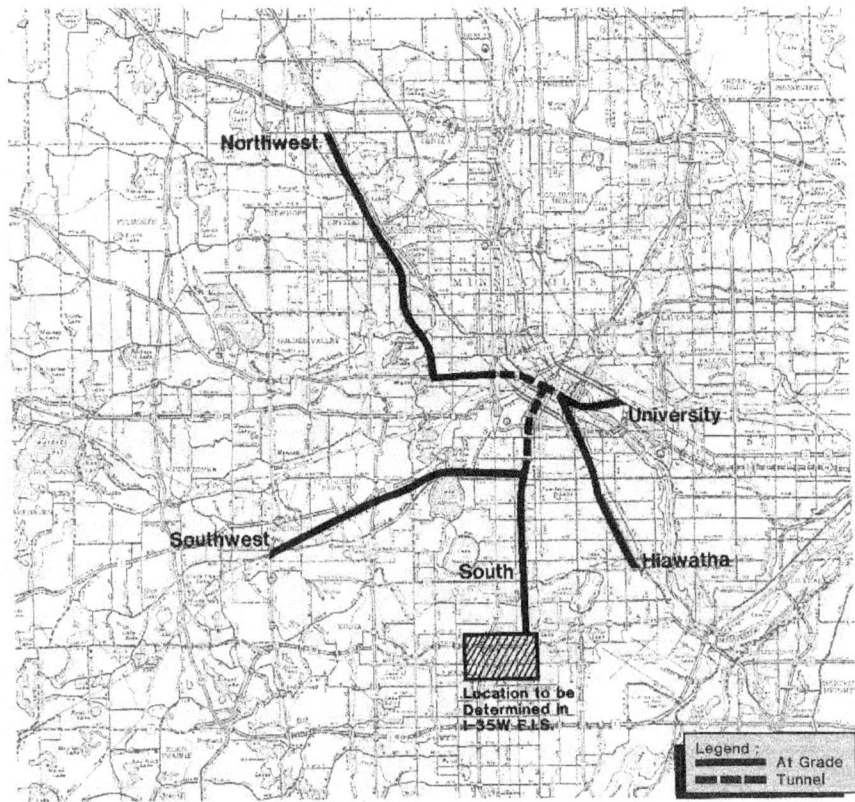

FIGURE 2 Stage 1 plan.

preliminary engineering, design, and construction" and also stated that the "funds appropriated for LRT should be considered as base level funding for presentation in the 1990–1991 biennial budget."

LRT PROJECT COMPONENTS

Figure 4 illustrates the major LRT system implementation components. LRT design and construction consists of the activities necessary to put the project's physical components in place: civil construction, procurement and installation of vehicles and their support systems, and construction of stations. The components include the following:

• Civil—the basic infrastructure of the system. For the purpose of simplification, this element involves preparation of the roadbed; all work below the subballast of the trackway, electrical subsystem foundations, underground conduit banks, drainage, subsurface treatment and grading; bridge structures;

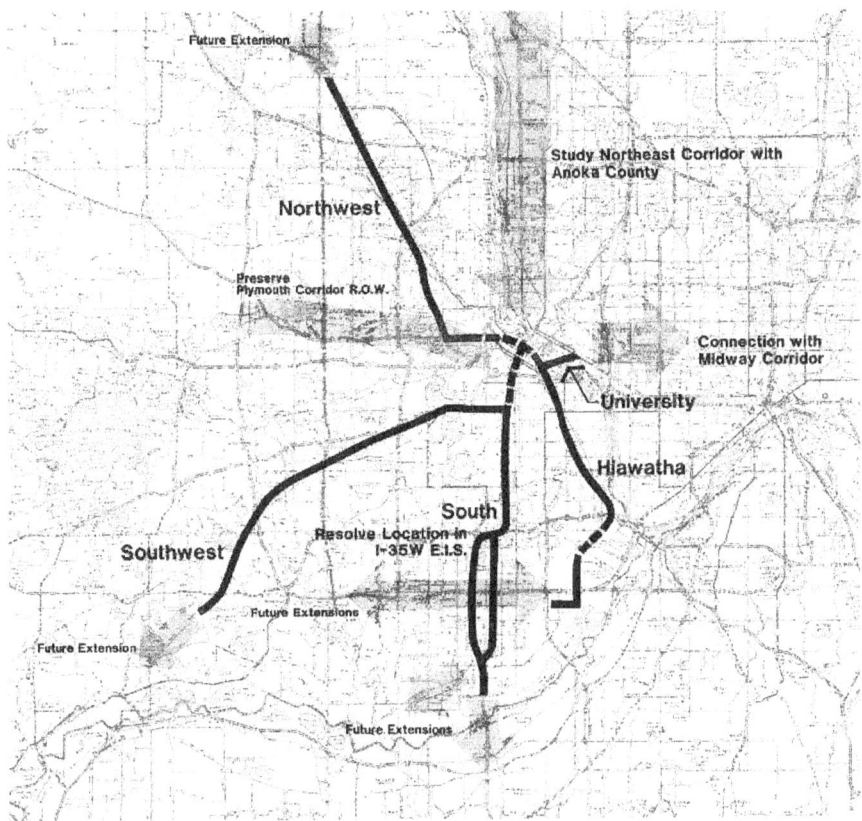

FIGURE 3 Twenty-year comprehensive LRT system plan.

street work; station footprints; and in the case of a subway section, tunnel construction.

• Systems—This element includes all facilities and equipment that are common throughout the system, i.e., light rail vehicles, track installation, electrification (power substations and overhead wires), signals, communication, fare collection, support equipment, and the central operations and maintenance facility.

• Stations—This involves station furnishings over and above the basic station footprint, including platforms and surface treatment finish work; lighting; furniture and amenities; electric power; shelters; heat; connections to roadways, public sidewalks and buildings; park-and-ride lots; elevators and escalators for any subway construction; and handicapped access.

Operations will commence upon the completion of design and construction. Public policy decisions (who will run the system and how) must be made prior to this event, an LRT operating structure defined and staff trained,

TABLE 1 CHARACTERISTICS OF RECOMMENDED 20-YEAR AND STAGE 1 PLANS

	TWENTY-YEAR PLAN			STAGE I PLAN		
SEGMENT	Length (Miles)	Capital Cost (1988 $ Million)	Daily Ridership Range Year 2010	Length (Miles)	Capital Cost (1988 $ Million)	Daily Ridership Range Year 2010
Downtown (Tunnel to 29th Street)	3.4	$138	--	3.4	$138	--
Northwest Corridor	12.0	139	19,600 - 25,500	9.0	114	18,000 - 23,500
Southwest Corridor	13.5	127	16,600 - 22,000	6.9	71	14,500 - 18,800
South Corridor	10.4	216	24,500 - 32,000	4.4	80	15,300 - 20,000
Hiawatha Corridor	10.0	145	17,300 - 22,500	3.9	34	13,000 - 17,000
University Connector	1.5	40	9,200 - 12,000	1.5	40	9,200 - 12,000
Yards and Shops	--	20	--	--	20	--
TOTAL	50.8	$825	87,200 - 114,000	29.1	$497	70,000 - 91,300

NOTE: The capital costs and patronage forecasts will be refined in Preliminary Engineering. The ridership forecasts are based on work reported in the Metropolitan Council report dated December 1986, "A Study of Potential Transit Capital Investments in Twin Cities Corridors" and the results of the Patronage Forecasting Peer Review Committee work.

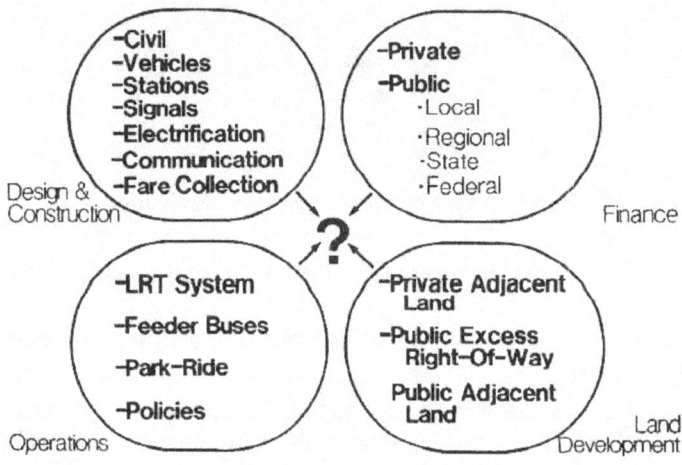

FIGURE 4 LRT project components.

and bus feeder services planned to coordinate service to the public. LRT will require that an organization be established, either within the structure of the existing transit agency or by a new operator, with rules and procedures appropriate to rail operations as distinct from the requirement of the present all-bus system. Personnel will be needed who possess specialized skills: transportation supervisory staff and train operators, security staff, vehicle maintenance personnel, and facilities maintainers.

Financing must be arranged from some combination of public and private sources to fund design and construction and ongoing operations. Related land development is likely to occur in the public right-of-way used by LRT as well as on adjacent private lands. Mechanisms can be implemented to capture for LRT system use a portion of the revenue that these new developments will create.

The issue addressed in this paper is how the above-defined LRT components should be related during system implementation and operation.

IMPLEMENTATION OPTIONS

LRT project implementation must efficiently coordinate the design, specification, procurement, and installation of equipment and construction of the LRT facilities. The objective of LRT project implementation is clear: on-time completion within budget with performance up to or exceeding the specification.

The major question is how to achieve this objective. To determine the best way to coordinate these facets of the implementation process, it is appropriate first to examine the various contracting methods as well as the roles and responsibilities of the implementers and then to match the contracting methods with the alternative implementation methods.

CONTRACTING METHODS

Several types of contracting methods may be used, each tailored to facilitate contractor performance of a particular set of construction, procurement, or furnishing and installation tasks. One-step competitive bidding (method A) is traditionally used when contract documents are clearly drawn and prospective contractors have a firm basis for their price proposals without significant latitude in interpretation. Advertisements solicit firm-price bids, after which award is made to the lowest responsive and responsible bidder.

Two-step competitive bidding (method B) is used when there is need to evaluate the bidders' approach to the project and their abilities to meet the stated objective. In these cases, the various prospective contractors have the latitude to approach the contract differently; and the owner reviews and selects the approach best suited to the original requirements before the contract award.

This process begins by advertising for technical proposals from potential contractors. Step 1 (which could be preceded by prequalifications, if desired) entails reviewing proposals (and possibly negotiating with the proposers separately to revise their technical proposals to meet the owner's needs). A limited number of responsive, responsible proposers judged to be capable of meeting the owner's needs are invited to submit prices. Step 2 makes the award to the lowest bidder.

Competitive negotiations (method C) are used when lowest price is not the only basis for award at the end of the evaluations. The process usually starts with an advertised request for letters of interest and qualifications. Proposals including technical approach and price are then requested from a screened list of qualified proposers. The main factors that are evaluated in the proposals are technical quality and price. Other factors, such as experience and performance history, may also be evaluated. Discussions and interviews are held separately with the proposers, as in method B, and continue until the proposers are asked to provide their best and final offers. Award is based on the highest-ranked proposals in terms of technical quality, price, and other prescribed factors.

ALTERNATIVE LRT IMPLEMENTATION METHODS

LRT system implementation will include not only the construction and procurement of system facilities and equipment, but also the financing of this work. In addition, options may be available to involve construction and procurement contractors in the operation and maintenance of the system after it is built. Recent years have also seen great interest in coordinating land development with rail transit construction. In some instances, developers of adjacent land have participated in the financing of transit stations.

The alternative methods of dividing the implementation work are discussed below. There are variations and hybrids of the methods shown, but those outlined constitute the basics for purposes of discussion.

Traditional

In the traditional method the project manager or engineer specifies the system elements (vehicles, electrification, signals, communications, fare collection, etc.) or components of the system elements (substation equipment, catenary network, track material, etc.) and issues separate detailed specifications for bid (see Figure 5). At the same time, the civil design is advanced to 100 percent drawings. Contracts are awarded for the system elements and components, and the contractors fabricate and furnish the equipment. The civil contract drawings are also issued for bid and awarded to low, responsible bidders; the contractors construct the LRT infrastructure. These construction contractors (or other contractors) could also install the electrification, signals, communication equipment, and fare collection. Upon completion, an operations contractor or a public agency operates the system.

Traditional contracting provides maximum control to the project owner, but limits the likelihood of obtaining contractor financial participation.

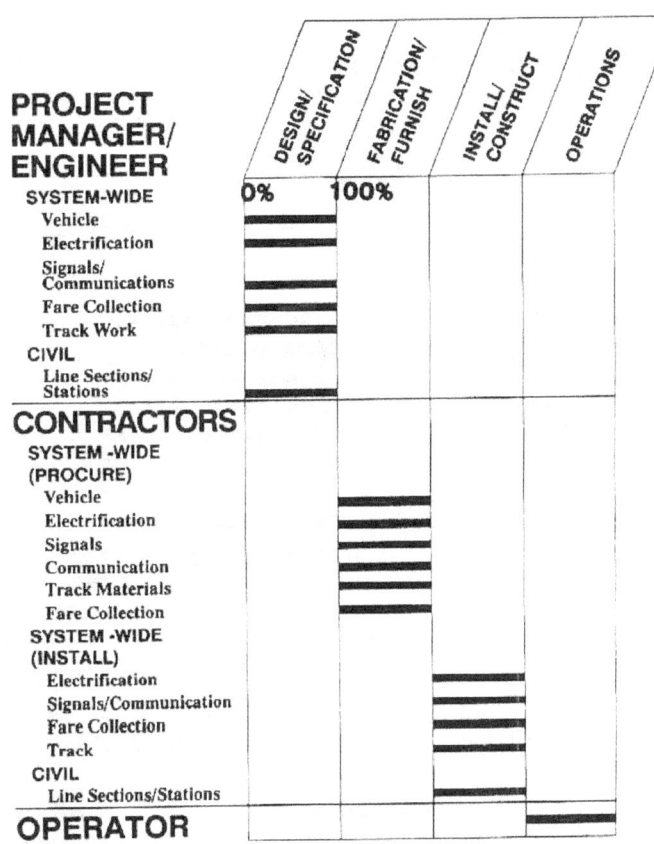

FIGURE 5 Traditional method.

Design/Build

In the design/build method, the project manager or engineer advances the design to the performance specification level in the case of the systems elements and to 30 percent in the case of the civil design (see Figure 6). The system elements each are awarded to contractors who design, furnish, and install the equipment. The 30 percent civil designs are issued for bid as design/build sections. Upon completion, an operations contractor or a public

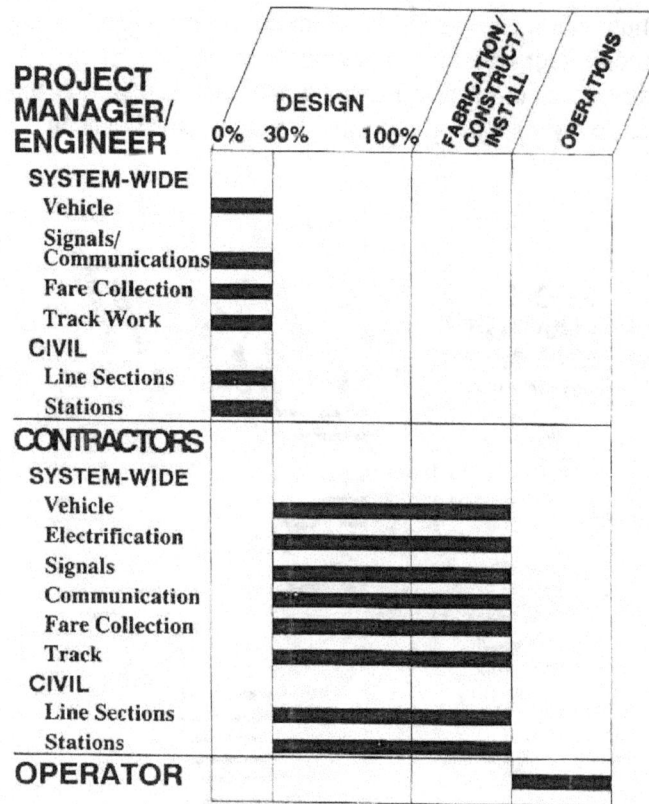

FIGURE 6 Design/build method.

agency operates the system; the operations decision is made independently of the design and construction.

The design/build method sacrifices a modest degree of owner control, but enables suppliers to tailor final design to their products rather than having to "reengineer" to the owner's exact specifications. Unless properly specified and managed, this approach can have the effect of limiting competition, thus affording an advantage in subsequent extensions to those firms successful in the initial stage.

Turnkey

In the turnkey method the project manager or engineer advances the design as would be done in the design/build method, but the performance specifications and 30 percent design are issued for competition as one package (see Figure 7). Having the project manager or engineer advance the design to 30 percent

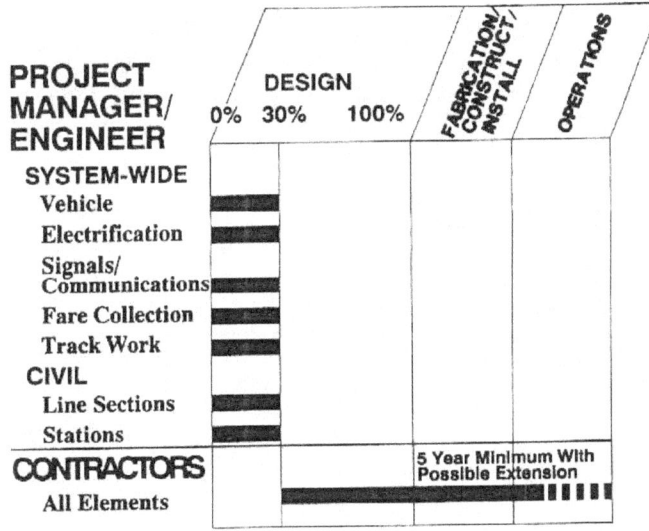

FIGURE 7 Turnkey method.

establishes the basic system parameters, allows for definitive cost estimation, and keeps the contingency margin reasonable.

The winning turnkey contractor completes the design in all areas and fabricates and furnishes the equipment at an agreed-upon price. The turnkey contractor also operates the system, at an agreed-upon price, for a prescribed period to ensure reliability. A minimum period of 5 years is usually suggested as a reasonable time period for problems to develop.

Turnkey further lessens owner control, but transfers responsibility for successful system operation to the turnkey contractor. Properly specified and managed, this approach focuses responsibility for cost and schedule performance, quality, and achievement of performance standards in a single entity. This removes many external interface-related claims.

Super Turnkey

The super turnkey method is the same as the turnkey approach except that the super turnkey contractor is also made responsible for partial or total system financing and is involved in the related land development. Financing might take the form of loans (e.g., vendor financing) or lease/buy-backs. There also could be a relationship of funding portions of the system, particularly at or around stations, through joint development.

The super turnkey approach makes the contractor responsible for financial and land development arrangements, but is likely to require that public agencies cede substantial control over the precise details of the technical and physical solution to the super turnkey contractor.

Contracting methods appropriate for each alternative implementation method are shown in Table 2.

TABLE 2 MATCHING ALTERNATIVE IMPLEMENTATION AND CONTRACTING METHODS

Alternative Implementation Method	Alternative Contracting Method
Traditional	One-step competitive bidding
Design/build	Two-step competitive bidding or competitive negotiations
Turnkey	Competitive negotiations
Super turnkey	Competitive negotiations

EVALUATION OF ALTERNATIVE IMPLEMENTATION METHODS

The evaluation of the applicability of the alternative implementation methods centers on the following criteria:

- Contractual, construction, and performance risk;
- Time schedule;
- Responsibility/accountability;
- Budget control/cost; and
- Quality.

Figure 8 presents a more detailed indication of how the various elements of LRT design and construction and operations fit together, and how they relate to options for public and private finance and development. Of 76 possible points of interaction, there are 32 "strong" and 25 "moderate" interrelationships.

These interrelations along with the above-defined criteria are used to reach conclusions. Although different metropolitan areas most likely will reach different answers about which implementation method to use, certain conclusions are reached on each of the implementation methods.

Civil

This element carries the greatest number of unknowns (e.g., soil condition variances), involves numerous third parties (utilities, railroads, and other

		LRT Design & Construction						Operations			Finance		Development	
		Vehicles	Civil	Stations	Communication	Signals	Electrification	LRT System	Feeder Bus	Public Policy	Public	Private	Public	Private
LRT Design & Construction	Vehicles													
	Civil	S												
	Stations	S	S											
	Communication	M	M	M										
	Signals	S	M	S	M									
	Electrification	S	M	M	M	S								
Operations	LRT System	S	S	S	M	S	M							
	Feeder Bus	N	M	S	N	N	N	S						
	Public Policy	S	S	S	N	M	M	S	S					
Finance	Public	M	M	S	M	M	M	S	M	S				
	Private	M	N	S	N	N	N	N	N	S	S			
Development	Public	N	M	S	N	N	N	S	M	S	S	M		
	Private	N	N	S	N	N	N	M	N	S	M	S	S	

Legend: S = Strong, M = Moderate, N = None

FIGURE 8 Relationships among LRT system implementation components.

public jurisdictions), and also involves property acquisition. If a subway or tunnel is part of the LRT, the risks are even greater. With a 30 percent level of design completed by the owner, all potential contractors have to either complete a significant amount of additional engineering or include a significant contingency in any fixed-price bid.

The failure to make right-of-way available or to gain agreements with railroads has been a historic problem and a cause of many schedule delays on fixed-guideway projects.

The schedule delays have also resulted in increased costs caused by inflation. The owner will have to take this risk and establish a firm schedule for availability of right-of-way and clearance of all utilities with a turnkey or super turnkey approach. It does not appear feasible to use a turnkey or super turnkey approach for all portions of the civil component of an LRT system.

The traditional method affords the highest degree of control. Civil design can be paced and adjusted in accordance with systemwide design development, third party negotiations, and the overall project schedule. The owner or the owner's project manager or engineer can fast-track certain long-lead sections (e.g., bridges) and adjust implementation schedules on other sections as the need arises.

Systems

Systems procurement for furnish/install contracts for several North American LRT projects (e.g., Portland, Sacramento, San Jose) has successfully been implemented using the design/build approach. Some foreign projects (Istanbul, Tunis, Manila) are using the turnkey approach.

An important consideration is the integration of the various systems components with each other and with the civil components. Most integration problems encountered will fall into two categories: systems/civil coordination and the securing of approvals and permits from regulatory bodies. The owner's project manager or engineer must possess the requisite skills to ensure this coordination. The systems integration function is crucial to ensuring that an operable project is built. Under the design/build option, the owner, through the project manager or engineer, could perform the coordination between civil and systems and can perform the coordination among the systems components (vehicles, signals, etc.) as well. This will allow tighter control by the owner.

Regarding the turnkey approach, no single manufacturer can provide all of the systems components (see Table 3). Thus, a turnkey approach will require several companies to cooperate, organized either as a joint venture or as a prime contractor with subcontractors.

Some suppliers have expressed an interest in the turnkey approach based on experience with projects outside the United States. The "price" of some loss of control by the owner may be worth considering if the turnkey contractor is prepared to accept some of the cost and schedule risks, and if the contractor is made responsible for operations management of the system for an extended period of time beyond the normal 2-year warranty time (say at least 5 years overall).

There is some thought among transit engineers that the contractual link between the major components of building the system and operating it may bring additional benefits. By holding the contractor responsible for management of operations (local forces already in place will perform actual works under the turnkey contractor's management), there is a financial incentive not to allow operating costs to exceed initial projections. The contractor may be more careful to design equipment to reduce operating and maintenance costs, because equipment failures will reduce the contractor's profit. Conversely, reliable, maintainable equipment will reduce costs, hence increasing profit.

In conclusion, a design/build approach for systems components will apply contracting methods successfully used on other recent North American LRT

TABLE 3 REPRESENTATIVE LIST OF LRT SYSTEMS MANUFACTURERS BY AREA(S) OF SPECIALIZATION

Company	Home Country	Light Rail Vehicles		Track	Traction Power		Signals/Control		Communications			Maintenance		Access to Financing
		Bodies	Mech/Elec	Matls	Substas	OH Equip	Rail	Traffic	Radios	Phones	CCTV	Tools	Vehicles	
Alsthom	France	X	X											Y
Siemens	Germany		Proplsn		X	X	X^a		X	X	X			Y
BBC	Switzerland		Proplsn		X					X				N
ASEA	Sweden		Proplsn		X									Y
NYAB	U.S.		Brakes											N
Ohio Brass	U.S.				X	X								N
GRS	U.S.							X						Y
WABCO	U.S.		Brakes				X							?
Bombardier	Canada	X												Y
Duewag	Germany	X												?
UTDC	Canada	X												Y
WECO	U.S.		X											Y
Central Power	U.S.				X									N
Hegenscheidt	Germany											X		N
Stanray	U.S.											X		N
Motorola	U.S.								X					N
Bethlehem Steel	U.S.			Rail										N
L.B. Foster	U.S.			Ties										N
Niedermeyer-Martin	U.S.			Ties										N
Sumitomo	Japan	X		Rail										Y
Breda	Italy	X												Y

^{a}Designs not compatible with U.S. practice.

projects and allow tighter owner control. A turnkey approach, however, may offer additional benefits regarding risk transfer and operating responsibility, albeit at the price of reduced county control.

Stations

The most appropriate implementation method for stations depends upon whether or not adjacent development opportunities exist. Traditional contracts are most appropriate to construct those stations where no developer involvement will occur. This will ensure maximum county control and coordination with other stations.

Super turnkey contracts are suggested where stations can be provided (i.e., built and paid for) by developers as part of adjacent building projects. Such contracts must be drawn to ensure compliance with LRT functional requirements (e.g., platform dimensions, weather protection for waiting passengers, station utilities, etc.); but some latitude may be given to allow developers to coordinate station architectural appearance with their projects.

Related Land Development

Counties in the State of Minnesota have no control over local land use decisions. They do not zone property and they do not approve building and site plans. Therefore, Hennepin County needs to establish interjurisdictional agreements with the various municipalities in which LRT service is proposed. With this completed, the county (as LRT developer) and municipalities can proceed together to solicit land developer or property owner interest and coordinate the development of stations integrated with adjacent real estate projects as discussed above and other developments on private land adjacent to the LRT right-of-way.

CONCLUSIONS

On the basis of the above discussion, two alternative approaches are suggested for metropolitan areas to pursue for the implementation of LRT systems.

Alternative A would use the traditional method for the civil component and the station/land development where no developer interest exists (see Figure 9). A design/build approach would be used for the system elements. Super turnkey would be used for station development where developer interest does exist. This approach retains significant control and responsibility with the owner, but allows the demonstrated advantages of design/build for the system

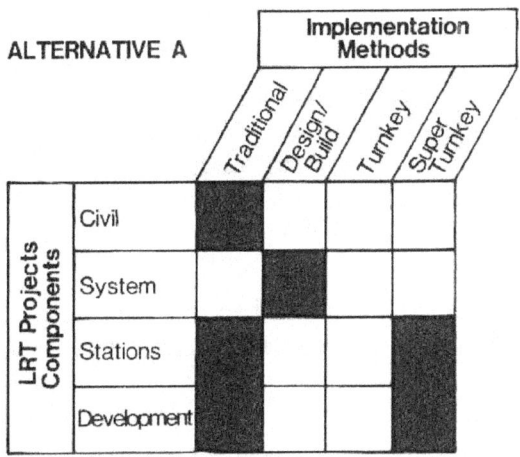

FIGURE 9 Alternative implementation methods for LRT project components (Alternative A).

elements. The role of the private sector relates to station construction and related land development. The objective would be to capture a portion of the increased value created by the presence of LRT.

Alternative B would result in maximum involvement of the private sector (see Figure 10). After the 30 percent design level the selected contractor would complete the design, build the system, operate the system, and be committed to capital contributions that relate to development around the stations or other innovative financing techniques. The contractor would be

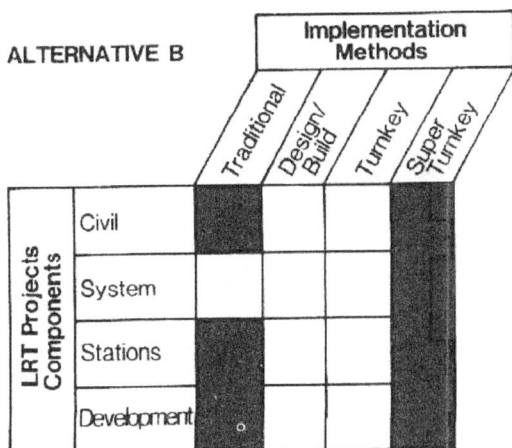

FIGURE 10 Alternative implementation methods for LRT project components (Alternative B).

selected on the following basis: financial capability, technical capability, management capability, and approach to "innovative" funding.

One of the problems with Alternative B is that no system has been completed using this approach. Many members of the private sector have recommended this approach, but to date it has not been used. One approach to giving Alternative B an opportunity without total commitment would be to follow the steps outlined below:

- Complete the preliminary engineering to the 30 percent level;
- While the engineering activities are being completed conduct the following tasks: prequalify super turnkey contractors and solicit technical approaches for innovative financing from the prequalified contractors; and
- If a proposal has substance, proceed with the super turnkey approach. If none of the prequalified proposers present "innovative" financing, continue with Alternative A.

West Side Manhattan Transitway Study

GREGORY P. BENZ, WENDY LEVENTER, FOSTER NICHOLS, AND
BENJAMIN D. PORTER

In response to the current and anticipated changes in the type and intensity of land use activities on Manhattan's West Side, the New York City Department of City Planning is conducting the West Side Transitway Study to ensure that adequate transportation services are in place to serve the new workers, residents, and visitors. The study, funded by the UMTA, is a four-phase effort that is examining potential transportation problems created by the anticipated developments, defining the degree to which improvements are needed, and determining the feasibility of implementing and operating new transit services and facilities to solve the identified problems. Because current sources of funding for public transportation are fully committed to the operation, rehabilitation, and upgrading of New York City's existing systems, innovative methods for financing and implementing the recommended improvements are being explored. This paper summarizes the first three phases of the study's transportation component. The existing transportation conditions in the study area are explored along with the future problems and needs created by the new development. The type of transportation improvement alternatives developed, primarily light rail transit (LRT) options, and specific issues related to reinstituting LRT in a dense urban environment such as Manhattan are described. In addition, issues related to privatization of the project implementation and operation are reviewed.

G. P. Benz and F. Nichols, Parsons Brinckerhoff Quade & Douglas, Inc., 250 W. 34th Street, New York, N.Y. 10119. W. Leventer, Manhattan Office, Department of City Planning, 2 Lafayette Street, Room 1400, New York, N.Y. 10007. B. D. Porter, Price Waterhouse, 1801 K Street, N.W., Washington, D.C. 20006.

TO COPE WITH SIGNIFICANT changes in land use activities on Manhattan's West Side, the New York City Department of City Planning is conducting a major study to examine potential transportation problems and to seek solutions. Funded by UMTA, the West Side Transitway Study will develop land use strategies and a financial, legal, and institutional plan to support any transportation improvements. The first phase of the study's transportation component identified the transportation problems and needs within the study area, shown in Figure 1. (Note: The study area indicated in the figures in this paper is the original designation in which the east boundary in Midtown Manhattan was Lexington Avenue; the study area was later extended further east to the East River.)

In the second phase, the initial set of alternatives was refined and evaluated in the light of physical and operational constraints, costs, and ability to serve new travel patterns and demand levels. A reduced set of alternatives was carried into the third phase where more specific travel demand analyses, engineering, cost estimates, environmental evaluations, and public policy analyses are used to select a preferred alternative. The last phase of the study will package the proposed transportation improvements with the results of the land use and financial and institutional evaluations to form an integrated strategy for addressing the future transportation needs of the West Side. This paper summarizes the first three phases of the study, including issues related to privatization of the project implementation and operation.

EXISTING CONDITIONS

The primary transit service on the West Side of Manhattan is two subway (rapid rail transit) lines in the north-south direction—the Eighth Avenue IND (Figure 1, A, C) and the Broadway-Seventh Avenue IRT (1, 2)—and three in the east-west direction—the 53rd Street IND (E, F), 42nd Street Shuttle and Flushing IRT (7), and the 14th Street (Canarsie) (L) lines. The major crosstown streets have bus services that generally run river to river. The north-south avenues as far west as 10th Avenue also have bus services. The study area also contains several major commuter facilities, including the Port Authority Bus Terminal, Penn Station, Grand Central Terminal, the Port Authority Trans-Hudson Corporation's (PATH's) Uptown (33rd Street) and Downtown (World Trade Center) terminals, the Staten Island Ferry Terminal, and the George Washington Bridge Bus Terminal.

The travel demand for most of these services and facilities is approaching or exceeds the supply at present. During the morning peak period the express services of the Seventh Avenue IRT and Eighth Avenue IND subway lines are overcrowded where they enter the hub of Manhattan, specifically the southbound services at 60th Street. These lines are also congested during the

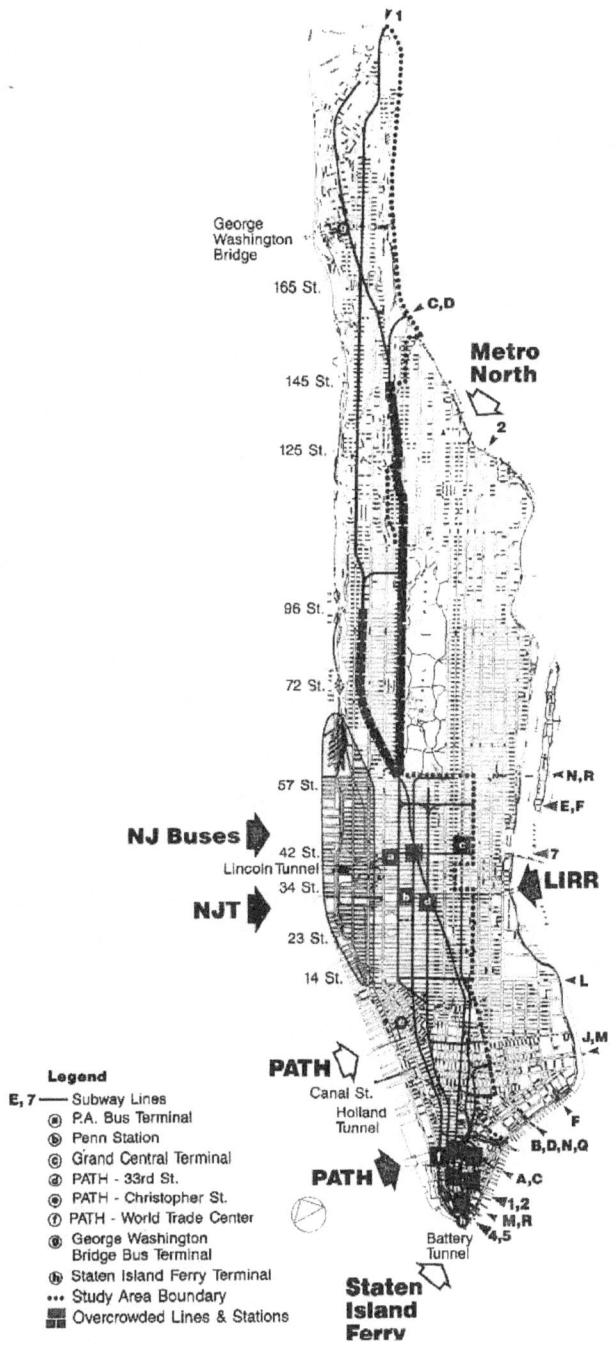

FIGURE 1 Existing conditions—transit.

evening peak, in the opposite direction. (The overcrowded conditions exist at current scheduled train service; however, the subway lines have the capacity to handle additional trains that could alleviate some of the overcrowded conditions.) Many of the subway stations in the study area are among the busiest in the subway system—Grand Central, Times Square, 34th Street (Penn Station), and Herald Square—and also experience significant crowding during peak periods.

Like much of the core of Manhattan, the West Side street network experiences increasing traffic congestion resulting from a growing reliance on travel by motor vehicles, particularly automobiles and taxis. Most of the north-south avenues and the major crosstown streets in the study area operate at congested levels—level of service D or worse on a scale of A (best) to F (worst). Travel speeds in the congested streets are as low as 2.5 mph, which is slower than average walking speed. Slow operating speeds and high traffic volumes combine to create concentrations (called "hot spots") of carbon monoxide pollution in excess of federally accepted air quality levels.

In addition to these problems, the area west of 9th and 10th avenues between 14th and 72nd streets has no subway service (see Figure 1). While this deficiency is not a major contributor to the problems with the transportation system today, the situation will change as this area is built up with new housing and offices.

FUTURE TRAVEL PATTERNS AND NEEDS

Recent and proposed development on the West Side is concentrated in four areas: Battery Park City/World Trade Center; Penn Station/Convention Center area; Lincoln Square West, which includes the proposed Trump City development; and Port Harlem at the west end of 125th Street. Most of this development is in the area with no subway service. Overall, the new residential development on the West Side will produce an estimated 22,000 new morning peak hour transit trips—a 10 percent increase over the number produced by today's residential inventory. The new office development will attract an additional 136,000 morning peak hour transit trips into the study area—a 14 percent increase. The evening peak hour will generate similar numbers, and the presence or expected construction of several major special trip generators—the Javits Convention Center, a new, relocated Madison Square Garden west of Penn Station, and a 1.5-million-ft^2 retail shopping mall in the Trump City proposal for the former rail freight yards between 60th and 72nd streets along the Hudson River—will create added loads on the transit system during the evening peak hour.

Most of the trips attracted by the new commercial development will originate in established residential areas of the outer boroughs and the surrounding suburbs. Many of these people will use the existing public transportation network to commute to midtown and downtown subway stations and commuter terminals. These transit patrons attracted by the new office space require connections between the existing transit system and the developing areas of the West Side not currently well served by the existing system. Specifically, the development concentrations in the area between 14th and 72nd streets west of 10th Avenue need to be tied into the stations of the Seventh Avenue and Eighth Avenue subway lines and possibly other lines farther east, as well as the Port Authority Bus Terminal, Penn Station, Grand Central Terminal, and PATH Terminals (see Figure 2). South of 14th Street and north of 72nd Street, existing subway lines are situated in proximity to the areas of proposed development and should be able to handle the anticipated transit trip levels.

Most of the trips from the new residential development will have destinations in the established employment centers in midtown and lower Manhattan, and will need direct transit links or connections to the existing transit system serving these established areas.

The majority of the trips generated by the new developments will end up on the existing transit lines and services. Although there is the potential that the 130,000 new peak hour transit trips—some 55,000 new trips in the peak hour on the Seventh Avenue IRT and Eighth Avenue IND subway lines alone—will increase crowding on the subway lines and stations, the capital programs of the New York Metropolitan Transportation Authority (MTA) and other transportation operating agencies in the region are designed to enable the existing system to handle these new trips.

The new West Side development will generate additional automobile and taxi trips on the already congested street network. In the absence of significant new transit connections in the area between 14th and 72nd streets west of 10th Avenue, 5,100 automobile and taxi trips in the morning rush hour and 7,500 afternoon peak hour vehicle trips will be generated. These numbers are up to twice the number of trips that would be generated if improved transit connections were provided. With a good transit connection available, this area would generate 3,400 fewer trips in the morning peak hour—a drop of 52 percent—and 2,800 fewer trips in the afternoon peak hour—a drop of nearly 30 percent. Given the problems of traffic congestion and air quality in the West Midtown area, it is essential that convenient, comfortable, and secure transit connections be provided to minimize the number of vehicle trips generated by the new developments.

The shift in the modal distribution of new trips away from automobiles and taxis to transit as a result of new transit connections will generate an estimated 7,300 additional transit trips in the morning peak hour for a total of 31,300 new transit trips in the peak hour (a 30 percent increase over existing

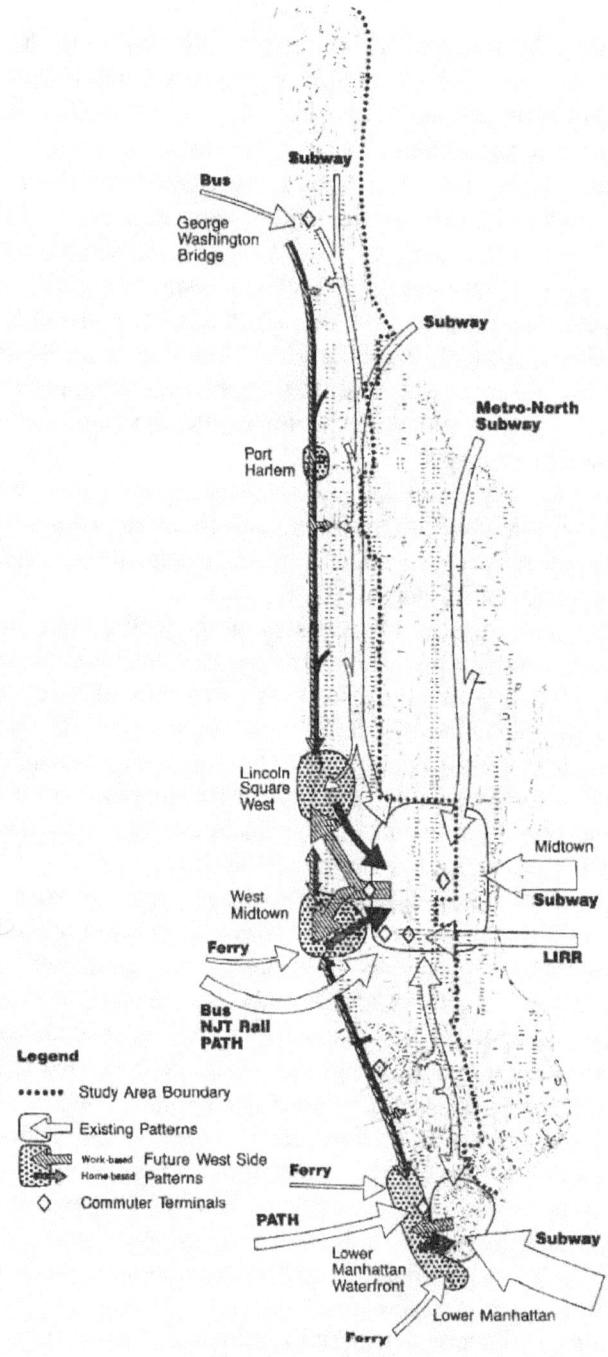

FIGURE 2 Transit demand patterns, morning peak.

transit volumes on the West Side). An additional 10,400 transit trips would be generated in the afternoon peak hour (a 28 percent increase) for a total of 47,000 new transit trips in the area between 14th and 72nd streets and west of 10th Avenue. The need to reduce traffic congestion and carbon monoxide produced by vehicles, as required by the policy of city, state, and federal governments, necessitates that good transit service connections be provided to the major development clusters on the West Side. This also will have the benefit of generating additional transit riders (and revenue) for the existing system.

In summary, the critical transportation needs identified in Phase I to be addressed by the West Side Transitway Study are twofold. First, to connect the developing areas of west Midtown between 14th and 72nd streets west of 10th Avenue with the Seventh Avenue IRT, Eighth Avenue IND, and other subway lines farther east; with the Port Authority Bus Terminal, Penn Station, PATH's 33rd Street Terminal and Grand Central Terminal; and to the midtown core. Second, to minimize automobile and taxi trips generated by the new development by attracting riders to transit through convenient, comfortable, and secure connections to the existing transportation system or to the midtown core.

In addition, the analysis of travel demand found that most of the trips from the new development were a combination of a north-south and an east-west trip and that a direct (no transfer) service was important in attracting riders to the transit service.

The West Side Transitway Study is focused on addressing the needs and problems of the developing area of the far West Side between 14th and 72nd streets, as well as considering the potential for improved transit services to the areas north and south. The problems of the existing public transportation systems and roadway network, such as crowding at stations and street congestion, are being addressed by the MTA, New York City and New York State departments of transportation, the Port Authority, and other state and local entities.

TRANSIT MODES

The most pressing transit need that emerged from the analysis of travel patterns was for a collector-distributor system that would connect the developing areas of the far West Side of Manhattan that are underserved by transit today with the existing commuter terminals and subway system in Midtown Manhattan. Extensions of the existing subway system, or new subway construction, are too expensive to be privately financed—the major thrust of this study. In addition, the spacing of the stations would not necessarily be compatible with the collector-distributor role of the proposed system.

Aerial structures across 42nd Street were determined to be unacceptable for environmental and aesthetic reasons. The lack of an affordable, completely exclusive right-of-way eliminated a fully automated guideway transit system.

Bus and light rail transit (LRT) emerged as the two modes of transportation that could work within the physical and operational constraints of the desired alignment. For many crosstown streets, improved bus service with some enhanced priority in the street to increase operating speeds meets the service needs. Along 42nd Street, however, the demand generated by either transit mode operating at the improved operating speeds offered by the transitway far exceeds the operational capabilities of bus technology. An unconstrained peak-hour, peak load point demand of 14,000 passengers is expected for a service operating at an average speed of about 9 mph at 3-min headways. As discussed later, LRT cannot handle the demand entirely.

PREFERRED ALTERNATIVES

In Phase III of the study, eight alternatives were analyzed, including no-build and transportation systems management alternatives. The six "build" alternatives were all LRT options that essentially addressed the problem to be solved in the area. The alternatives varied in the north-south direction using either a railroad right-of-way (ROW) (called the Amtrak Cut), 11th Avenue, or the West Side Highway-12th Avenue, or some combination. All the options included a 42nd Street crosstown segment, except one that went across 34th Street. All included a grade-separated transit link between Penn Station and the Long Island Railroad yard development site.

The primary difference among the alternatives was the capital cost and, to a lesser degree, the revenue generated. As a result of the financial analysis (discussed later), only one alternative emerged as being financially feasible under a viable privatization scenario. This alternative is a two-track LRT line across 42nd Street (river-to-river) in an at-grade transitway (Figure 3). On the east end is a loop track with an extra layover track. At the west end the line has two possible alignments: down 11th Avenue at grade to 30th Street, where it goes onto an aerial structure along 30th Street to Ninth Avenue and then eastward along 31st Street to Penn Station; or down the reconstructed West Side Highway (12th Avenue) in the median or along the western edge of the highway to 33rd Street, where it would go into the tunnel under 33rd Street to Penn Station at Eighth Avenue. In either case, the end of the line will be a two-track stub-end terminal. The resolution of the issues that will determine which of these alignments will be selected will not occur within

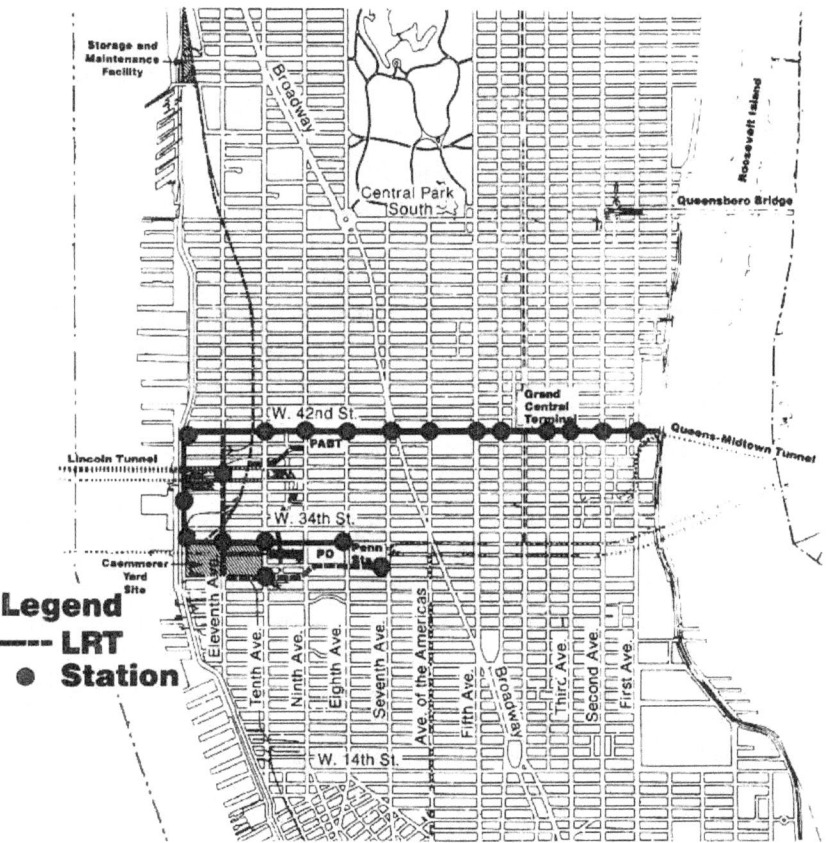

FIGURE 3 Preferred alternatives.

the time frame of the study, particularly the integration of the transitway with the highway reconstruction plans.

The peak-hour ridership for the system is forecast at 15,000 for the year 2005. The peak-load point demand of 14,000 (eastbound on 42nd Street in the area of Times Square) is constrained by the 10,000-passenger directional capacity of the system (two-car trains at 3-min headways). The total unconstrained peak hour demand for the system is estimated to be 19,000 riders. Even with the constraint, the daily ridership is expected to be 103,000 passengers. The annual ridership is forecast at 28.6 million, including over 3 million trips from the special trip generators such as Madison Square Garden and the Convention Center.

The operations are constrained by several factors. The block lengths in the north-south direction as well as some other factors limit the train length to under 200 ft. The study assumed two-car trains consisting of double-ended articulated vehicles of approximately 85 ft. The stub-end terminal at Penn

Station, heavy passenger boardings at several stops, and the crossing of all the major avenues along 42nd Street limited the headway to 3 min. While a shorter headway may be technically feasible, the 3-min headway was felt to be one that could be operated reliably and was used for planning purposes.

The vehicle for this service would have to have several features in order to handle the projected passenger loads and minimize dwell times. The internal configuration of the vehicle would have very few seats to allow maximum standing area. A total capacity (standing and seated) of 250 passengers per car is needed. To minimize dwell times, low-floor vehicles similar to the ones operated in Grenoble, France, or Geneva, Switzerland, are needed to facilitate loading and unloading from the curbside. High-level platforms are not feasible along 42nd Street.

An operational factor affecting dwell times is fare collection. Conventional on-board fare collection will not work at the high-volume stations; dwell time would be excessive. The issue still needs to be resolved, but an off-vehicle system that allows usage of all vehicle doors for loading and unloading is needed. Fare-controlled platforms are not feasible along 42nd Street. Hence, a self-service fare system, with inspection upon boarding at high-volume stations, is proposed.

All the alternatives, except the preferred option, have direct connections from the revenue tracks into the proposed vehicle storage and maintenance facility north of 72nd Street. The preferred alternative has no revenue service north of 42nd Street, although a future stage of system development could extend service farther north or south.

No viable sites for a full yard and shops exist adjacent to the revenue tracks south of 42nd Street. The 72nd Street site is the only location for the major maintenance and repair facility. The connection to the yards from the revenue tracks would be by way of the Amtrak Cut. The cut has sufficient width to have several lay-up tracks, but not a separate light transit connection to the yards. This connection would be over the proposed Amtrak tracks. Trains of light rail vehicles would be pulled over the Amtrak tracks by a diesel locomotive up to the yards. Overnight storage and major inspections and repairs would be done at the 72nd Street yards. Midday storage, running repairs, and daily inspections would be conducted south of the Amtrak connection in the railroad cut around 34th Street. While this arrangement is inefficient and imposes some potential operational constraints, it is the only means developed thus far to provide a yard and shop facility for the preferred alternative; without this, the alternative is infeasible.

The capital cost for the 3-mi line is $284 million. The factors contributing to the high capital cost include:

- Relocation of the maze of utilities under Manhattan streets;

- Maintenance of traffic for Manhattan's heavily traveled roadways;
- Large fleet requirements (35 vehicles) relative to the system length;
- Remote location of a maintenance facility; and
- Construction of the aerial structure or tunnel for the connection into Penn Station.

The annual operating and maintenance cost is estimated to be $7.5 million. This estimate includes costs for the fare collection system, the special operations for the connection to the remote yards, and extra service to handle the special events at Madison Square Garden and the Convention Center.

The ridership estimates are based on charging a separate $1 fare to use the system—distinct from any fare charged to use the existing transit system. A significant portion (two-thirds) of the expected morning peak riders are traveling along 42nd Street for the final portion of their journey from the commuter terminals or the subway system. In the absence of the transitway, these passengers would either use the crosstown bus (which will be replaced by the LRT line) or walk. The transitway offers a substantial increase in speed and capacity over the existing bus service. It is the combination of attracting both trips along 42nd Street and trips from the developing areas of the far West Side that provides the revenue that is the basis of the financing plan for the system.

FINANCIAL AND INSTITUTIONAL ISSUES

A major challenge in establishing the feasibility of the West Side Transitway was to determine whether the project could be constructed and operated without infringing on the region's ability to revitalize the existing public transportation system. This challenge posed two key questions. First, could the project succeed financially without relying on any of the revenue sources currently used to fund transit in the city? Second, under what type of institutional arrangements could these financial plans be implemented? An underlying theme to both of these questions, and an object of major interest to the study, was the extent to which private sector participation could expedite the project's implementation.

The financial and institutional analysis addressed these questions through the following steps. First, three financial plans were developed for each of the transitway alternatives. Each plan was intended to offer a different allocation of risk between the public and private sectors—100 percent private risk, shared public-private risk, and 100 percent public risk. Second, the feasibility of the plans was evaluated according to the internal rate of return to private investors in the project, and the amount of publicly derived revenues needed to make up the shortfall between operating revenues and the full cost of the project. This included a review of new revenue sources that could be implemented to support the project's cost. And finally, a review of the ability of potential public sponsors for the project was undertaken to determine the legal and legislative requirements to implement the project.

These analyses found that private sector financing of the project was more attractive than a traditional, tax-exempt financing approach. A special assessment was selected as the preferred revenue source. Further, the preferred alternative (described in the previous section) was the only transitway alternative that met the tests of feasibility, and the project could be implemented either by the city or by an existing or new state-created public authority. The key findings from each of these analyses are summarized below.

Financing Plans

The financing plans used in the analysis reflected a concept developed early in the study regarding possible approaches to the ownership of the transitway assets—a continuum ranging from complete private ownership to complete public ownership, with various options for shared ownership. The ownership of assets was stratified in this way because ownership dictates the types of financing mechanisms that can be used. The development of financing plans that reflected this distribution of ownership allowed the study to consider the merits of different financing mechanisms.

1. *Private Structure.* In the private structure approach, the transitway would be implemented via a franchise wherein a private company would assume all risks for the project. This approach relied on debt financing for construction and a combination of equity, senior lien debt, and subordinated debt for all permanent financing exclusive of vehicles. A leveraged lease was used for rolling stock procurement. Farebox revenues and advertising fees were the only sources of revenue for the project.

2. *Public-Private Structure.* In the public-private structure approach, the transitway would be implemented via a service contract between a sponsoring public agency and a private company or consortium. The private company would assume all risks for the cost of the project, while the public agency would assume all revenue risks. Thus, the public sponsor would agree to pay a negotiated annual service fee to the private company, irrespective of whether the operating revenues were sufficient to cover the fee. Revenue shortfalls would have to be made up from an alternate source. The financing structure used for the private company was similar to that used in the franchise approach above.

3. *Public Structure.* In the public structure approach, tax-exempt debt would be used to pay for construction and vehicle acquisition. Its chief

difference from the traditional approach to financing public transit capital projects is that no government grants were assumed to be available, and that the revenues (operating revenues and alternative revenue sources, such as special assessment) used to pay debt service cost would not be available until the transitway was operational. As a result, the capitalized interest costs would be substantial.

The results of this analysis contained some surprises. First, the public structure approach did not fare well because of the extent of capitalized interest costs. That is, the additional interest costs associated with 100 percent debt financing exceeded the benefits of the lower interest rate available through tax-exempt bonds. This finding is interesting in that it reveals the true cost of transit capital projects that is often masked when extensive federal financing is available. Second, the pretax rates of return for the private structure (i.e., the franchise approach) ranged from 4 to 12 percent—not high enough to attract investors. Given that the transitway alternatives are located in one of the most densely developed and transit-dependent areas of the country, these findings suggest that private sector ownership of capital-intensive transit systems is not viable without some public sector support. Finally, the annual shortfalls between full costs (operating and maintenance cost plus return on investment) and operating revenues for the public and public-private structures required that a strong and predictable alternative revenue source be available at least through the early years of the project.

Revenue Sources

Existing transit services in New York City are funded by a combination of operating revenues, bridge and tunnel toll revenues, general funds of the city, a mortgage tax, and grants from the State of New York and the federal government. All of these revenue sources were considered to be off limits to the project, given the intense and regionwide interest in revitalizing the existing transit infrastructure. Accordingly, the search for potential revenue sources focused on new mechanisms not needed to support the revitalization efforts.

The revenue sources considered in the analysis were all related in some fashion to the real estate development projected to occur in the study area. The rationale for the use of these revenue sources reflected two attributes of the transitway project: its ability to improve accessibility for travel to and within the service area, which should contribute to higher land values and rents; and its ability to mitigate the impacts to the existing transportation infrastructure associated with higher density development. It was generally agreed that existence of these benefits was essential to the acceptance and use of new, real estate-related revenue sources.

Six types of revenue sources were investigated:

- Special assessments—a fee (exclusive of property taxes) levied on property that is benefited by an adjacent or nearby public improvement. Special assessments have been used to support the financing of public transit improvements in Miami, Los Angeles, and Denver and were contemplated by the New York State legislature in the Rapid Transit Law of 1898.
- Tax-increment financing (TIF)—the dedication of incremental property taxes (above the current tax base) in a specified district to the financing of public improvements in that district. Although TIF is not commonly used to finance transit improvements, it was used to support the financing of the Embarcadero Station in San Francisco.
- Sale or lease of public property or air rights—the sale or lease of development rights above or adjacent to the station. It has been used as a source of revenue by the rapid transit systems in Washington, D.C., and Miami.
- Zoning incentives—incentives such as increases in the allowable floor-to-area ratio of a lot have been awarded by the city in return for the provision of certain public improvements (e.g., subway station improvements) by a developer, where these improvements are rationally related to the incentive being offered.
- Mitigation—actions taken by a person, or by a business entity, to minimize or avoid adverse environmental impacts associated with an action (e.g., a development) under consideration by a governmental agency. Developers have often provided public improvements as a component of the mitigative actions associated with new development (e.g., an esplanade along the East River was rebuilt by developers to mitigate an adverse environmental impact on open space).
- Impact fees—fees that are levied on new development and represent the new development's pro rata share of necessary public improvements that, but for the new development, would not be required to service existing residents. Although not commonly used for transit purposes, impact fees are levied on new office development in downtown San Francisco to support expansion of peak period transit services in connection with the increased transit demand generated by new office space.

Each of these potential revenue sources was evaluated with respect to five criteria: income generating ability (yield and profile), risk, legislative requirements, litigation risk, and administrative requirements. Special assessments were found to be the most logical choice. Although state enabling legislation would be required, this revenue source was preferred because: (1) it was established that the transitway alternatives enhanced accessibility (as measured by travel time savings) both to and within the study area; (2) these benefits would accrue to existing and to new development; (3) assessments could be collected concurrent with the city's collection of property taxes and be subject to the same system of remedies if collections were delinquent; (4) there exists legislative precedent within the state for its use; and (5) it could easily meet the revenue shortfalls projected in the financing plans.

Feasibility Analysis

Two criteria were used to evaluate the feasibility of the transitway alternatives. First, the pretax internal rate of return was used to evaluate the private structure finance plan. A minimum rate of 15 percent was believed to be necessary to attract investors to the project.

Second, the amount and duration of special assessments (based on the assessment per square foot of commercial property within walking distance of the transitway) was used to evaluate the public-private structure and the public structure finance plans. A rate of 30 cents/ft^2 was used as the threshold value. This rate was the approximate midpoint of the range of assessment rates in use in Miami and Los Angeles. These rates were adjusted to a comparable rate for New York City by normalizing for prevailing rents. This approach was used to ensure that the threshold rate was not so high as to deter new development.

The feasibility analysis found that only one of the transitway alternatives was financially viable and only under the public-private finance plan. This preferred alternative consists of an LRT line on 42nd Street between First and 12th avenues, on 11th or 12th Avenue between 42nd Street and the vicinity of the Jacob Javits Convention Center, and between the Convention Center and Penn Station. This alternative has a 15.4 percent internal rate of return and an assessment rate of 18 cents/ft^2 in 1994, declining to 3 cents/ft^2 in 1999 (the last year of the assessment).

It is notable that these results reflect relatively conservative assumptions on inflation rates [approximately 6.5 percent annually for construction and operating and maintenance (O&M) costs], financing charges, and ridership growth. Also, O&M and construction costs were modeled based on public sector experience. Nationally, private sector construction costs are approximately 15 percent lower. This would reduce the assessment rate by almost 50 percent and bring the private structure (i.e., franchise) approach to the threshold of feasibility.

Legal and Legislative Requirements

The financial analyses found that the preferred transitway alternative should be implemented through the use of a service contract between a sponsoring public entity and a private company or consortium. It is likely that the service fee to be paid to this company could not be borne by operating revenues alone, at least in the early years of the project—an alternate revenue source will be needed. A special assessment was found to be the most logical choice to provide these additional revenues.

The provision of transit services in the city via a service contract, and the use of a special assessment to support the funding requirements of these services, is a significant departure from the existing institutional environment. Currently, transit services are provided by the New York City Transit Authority (TA). The TA accordingly has all the powers necessary to operate transit service and to use city streets for this purpose. However, the TA does not purchase transit services via contract and its ability to do so on the scale envisioned for this project is open to question. Also, while the power to levy special assessments was apparently conferred on the TA in its enabling legislation (when it was conferred powers that were originally conferred on the city by the Rapid Transit Law of 1898), its ability to exercise this power has never been established.

Accordingly, an analysis of the legal and legislative requirements for implementing the project was conducted. This analysis consisted of a review of the requirements for establishing special assessment districts and a review of the powers of existing public institutions to implement the project using a service contract.

While the project's implementation by any public entity would require state legislation, the city may face the lowest hurdles. With the passage of a local law, the city could enter into contracts for the purchase of transit services. State enabling legislation would be required, however, to implement a special assessment district. There is no associated requirement for local approval of the special assessment.

A public authority, such as the TA or any other state-created public authority, could also implement the project, but not without additional state legislation and not without the city's involvement. For these authorities, state legislation would be required for at least the use of a service contract (for the TA), and possibly other mass transportation-related powers (if an authority other than the TA were to sponsor the project). State legislation would also be required for the special assessments, and this legislation would stipulate the city's involvement in the exercise of this power (e.g., in establishing the assessment rate).

CONCLUSION

This study generated several important findings that are relevant to consideration of LRT in comparable situations. The transportation need of new concentrations of dense development in urban centers is for a collector-distributor service connection to the existing transportation system. As such, frequent and easily accessible stations or stops are needed.

Placing a new LRT system into a very densely developed area that has limited feasible alignment options can introduce limitations on the operational potential (speeds, capacity) of the technology. The capital cost of constructing LRT is greatly affected by the environment into which it is placed. Relocation of dense old utilities, maintenance of traffic, and limited construction space can increase the cost significantly.

Given that the proposed transitway alternative is located in one of the most densely developed and transit-dependent areas of the country, this study indicates that private sector ownership of capital-intensive transit systems is not viable without some public sector segment.

Boston's Light Rail Transit Prepares for the Next Hundred Years

James D. McCarthy

For over a century light rail transit (LRT) has played an important part in the development of the City of Boston and its suburbs by fulfilling its transportation needs. Today, LRT runs over many of the same routes it did a century ago. As we approach the century mark of Boston's first electric trolley, it is appropriate to review some of the accomplishments of light rail in Boston and to look at the future. The Massachusetts Bay Transportation Authority (MBTA) has two light rail projects currently in design. A third proposal would extend the light rail system in the future. At North Station, the Green Line (light rail) will be relocated to a new subway alignment that will create a new transportation center. At Lechmere Square in Cambridge, the existing Lechmere Station will be relocated across O'Brien Highway to a new site that will enable the MBTA to develop a new station and a light rail vehicle maintenance facility. The relocated Lechmere Station is the first phase of a plan to extend the Green Line beyond Lechmere into Somerville and Medford.

MASSACHUSETTS BAY TRANSPORTATION AUTHORITY (MBTA) was created in 1964 as a political subdivision of the Commonwealth of Massachusetts to replace the Metropolitan Transit Authority. The MBTA has the responsibility of providing public transportation within the City of Boston as well as the surrounding 78 communities that make up the Regional Transportation District. The population of the 1,038-mi^2 district exceeds 2.6 million. The MBTA's net deficit after revenue and federal operating assistance comes from two sources: 50 percent from regional property tax assessments receipts and 50 percent from general state revenues.

Massachusetts Bay Transportation Authority, 10 Park Plaza, Boston, Mass. 02116.

The MBTA's system handles 600,000 passengers each weekday, employing 786 peak buses, operating over 150 routes covering 710 route mi; 4 light rail routes and 3 rapid transit routes operating on 183 mi of track; 4 trackless trolley routes covering 16 route mi; and a commuter rail system covering 357 route mi. The three rapid transit routes are distinguished as the Blue, Orange, and Red lines. The four-branch light rail system is known as the Green Line. The commuter rail system is the Purple Line (see Figure 1).

FIGURE 1 Boston transit system map.

HISTORICAL BACKGROUND

On January 1, 1889, the first electric trolley left the Allston Depot down Harvard Street to Beacon Street, traveling to its destination at Scollay Square in downtown Boston. As we approach the century mark of the first electric trolley to operate in Boston, it is appropriate to review the accomplishments of Boston's light rail system and look to its future.

This historic event had its origins in the first streetcar operation in the Boston region. On March 26, 1856, the Cambridge Horse Railroad, which had been organized in 1853 as the first street railway company in Massachusetts, inaugurated its first route, which ran from Harvard Square in Cambridge over Massachusetts Avenue, Main Street, and the West Boston Bridge to Bowdoin Square.

Not quite 33 years later, Boston's first electric car began operating from Allston to Scollay Square. The second electric line opened along Beacon Street less than two weeks later on January 12, 1889, running from what is now Reservoir Station at Cleveland Circle to Park Square. The third line opened the following day from Oak Square in Brighton to Park Square. By April 2, 1894, when the Boston Elevated Railway Company was chartered by the Massachusetts General Court, most of the streetcar lines were electrified and for the most part were still operating in the streets.

America's first subway was opened in Boston on September 1, 1897, when electric car No. 1752 from Allston entered the tunnel. Also in 1897, the Boston Elevated Railway Company took over the West End Street Railway. On September 3, 1898, the Tremont Street subway was extended from Park Street north to Causeway Street (North Station). There was a station at Scollay Square with the northbound side called Corn Hill and the southbound side, Tremont Row. The ensuing years saw the Boston Elevated Railway Company rapidly expand service, building the East Cambridge Viaduct to Lechmere that opened on June 1, 1912.

The Boston "El" was succeeded by the Metropolitan Transit Authority (MTA), and the MTA acquired the Boston & Albany Railroad from New York Central on June 24, 1958. On July 1, 1959, streetcar service was inaugurated on this new line into Brookline and Newton where the Riverside terminal is located.

Since August 4, 1964, when the MBTA succeeded the MTA, many improvements have been made to the Green Line. These include the modernization of Arlington, Government Center, Haymarket, Copley, Prudential, Kenmore, Auditorium (formerly Massachusetts Avenue), and Park Street stations; reconstruction of the Highland Branch (Riverside Line) by installing new roadbed and all-welded rail; and improving station platforms and lighting. In addition, new traction power and new signaling and communications equipment have been installed on the Riverside Line and in the Central Subway and a new track structure has been installed in the Central Subway.

TODAY'S LRT SYSTEM

The 27 mi of the Green Line (5 subway, 21 surface, and 1 mi elevated) and the 2.5 mi of the Mattapan-Ashmont branch of the Red Line are the last of the network of trolley tracks that once covered Boston and many of its suburbs. The Green Line runs on an elevated track from Lechmere Station in Cambridge to North Station in Boston, where it goes into the subway for Haymarket and Kenmore. The Central Subway provides connections to the three rapid transit lines—to the Red Line at Park Street Station, to the Blue Line at Government Center Station, and to the Orange Line at Haymarket Station (see Figure 2).

Kenmore Station in Boston's Back Bay is the last subway station before the line branches off for Commonwealth Avenue to Boston College in Newton; Beacon Street to Cleveland Circle through Brookline; and the Riverside rail right-of-way through Brookline and Newton to Riverside Station near Route 128 and the Weston line. The Arborway Line branches off at Copley Square, continues underground to Symphony, and then runs on the street to the Arborway in Jamaica Plain.

Operations

The President's Conference Committee (PCC) cars no longer run on the Green Line; they have given way to the new light rail vehicles (LRVs). The Beacon Street, Commonwealth Avenue, and Huntington Avenue lines still exist today almost as they did a century ago. The Central Subway is unchanged with the exception of station modernization and facility improvements. The Green Line carries approximately 220,000 daily riders and is the spinal cord of the MBTA's transportation system.

There are 56 colleges and universities in the Boston area and one out of every 40 college students in the United States attends classes here. The Green Line has direct service to several of these institutions: Boston College, Harvard Medical, Boston University, Northeastern University, Emerson College, Massachusetts College of Art, and Wentworth Institute of Technology. Also, Boston is blessed with some of the finest medical institutions in the world. Education and medicine provide one of every six jobs in Boston. The Green Line serves many of these hospitals.

Because the colleges and hospitals are located outside the central business district (CBD), they give the Green Line the unique quality of a two-way ridership demand during the peak and off peak hours.

Ridership

Over the past 20 years the MBTA has made major improvements to its rapid transit system. Major extensions and upgrades have occurred on the Red and Orange lines and the Blue Line has received new vehicles and track structure.

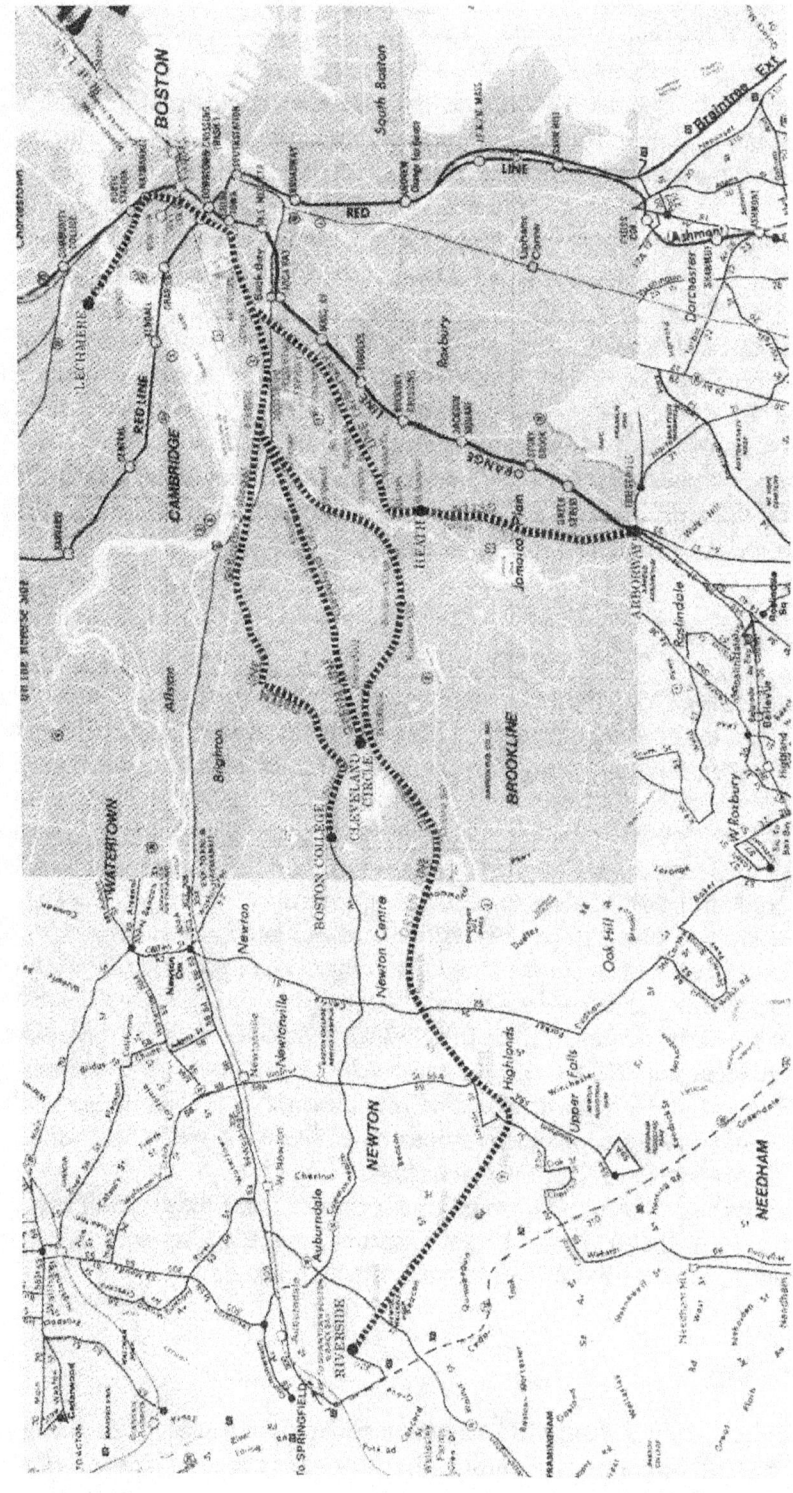

FIGURE 2 Green Line route map for Boston and its suburbs.

Demands for better transportation exist more today than ever. Ridership has increased on all lines, but the Green Line has experienced the most dramatic growth, with the usual consequences of operating at capacity. Although the other rapid transit lines have increased their capacity by adding cars to make longer train consists, the Green Line has been restricted by equipment problems, subway design, and a lack of LRVs to maintain an increased schedule.

Figure 3 shows the inbound surface ridership on the Green Line for all branches. Ridership has been on the increase for the past 10 years and indications are that it will soon pass the 25-year high. Of the 455,000 passengers/day that use the entire rapid transit and light rail system, approximately 220,000 include a Green Line segment. Of the total daily Green Line passengers, 39 percent make trips involving only the subway, and 17 percent make trips involving only surface segments. Table 1 breaks down the surface ridership of the Green Line. The figures for the Boston College line show that 35 percent of the total ridership is for surface only, indicating the strong student ridership for Boston University and Boston College.

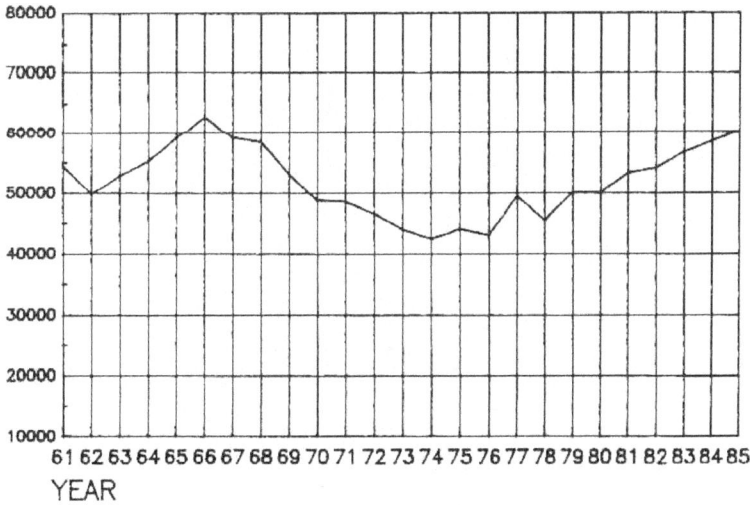

FIGURE 3 Light rail ridership, surface inbound.

Schedule

To meet the ever increasing demands on the Green Line, MBTA has developed two operating plans for future service levels—a 1990 service of 147 peak cars and a post-1990 service of 159 peak cars. Existing peak service is 125 cars.

TABLE 1 COMPARISONS OF GREEN LINE SURFACE TRIP GENERATION (7 a.m.–10 p.m.)

	Boston College (3.95 mi)[a]		Cleveland Circle (2.24 mi)		Riverside (9.25 mi)		Arborway Heath (3.6 mi)		All Branches	
	Total	Per Mile	Total	Per Mile	Total	Per Mile	Total	Per Mile	Total	Per Mile
In ons	15,837	4,009	9,310	4,159	13,729	1,484	16,153	4,487	55,029	2,888
Out offs	19,422	4,917	9,646	4,306	14,003	1,514	17,146	4,763	60,217	3,161
Two-way ridership	35,259	8,926	18,956	8,461	27,732	2,998	33,299	9,250	115,246	6,049
Inbound surface-subway	11,594	2,935	7,403	3,305	10,861	1,174	11,563	3,212	41,421	2,174
Inbound surface-only	4,243	1,074	1,907	850	2,868	310	4,590	1,275	13,608	714
Outbound subway-surface	11,196	2,834	6,624	2,957	9,442	1,021	10,793	2,998	38,055	1,998
Outbound surface-only	8,226	2,083	3,022	1,349	4,561	493	6,353	1,765	22,162	1,163
Two-way-surface-subway	22,790	5,770	14,027	6,263	20,303	2,195	22,356	6,210	79,476	4,172
Two-way-surface-only	12,469	3,157	4,926	2,199	7,429	803	10,493	2,915	35,770	1,878
Percent surface-only	35.4		26.0		26.7		31.5		31.0	

NOTE: 1985 counts.
[a]Surface length.

The post-1990 service will add cars to the 1990 schedule and possibly extend the Green Line beyond Lechmere. The impact of the proposed increased service levels will be discussed later in the context of the plans for the Lechmere Maintenance Facility and the extension beyond Lechmere. Table 2 shows the existing and projected Green Line service.

The following sections discuss how America's oldest subway system is preparing for the next hundred years.

PLANNED IMPROVEMENTS

In 1980 the MBTA undertook a study to examine the alternatives for making transportation improvements in the Green Line Northwest Corridor. The Green Line Northwest Corridor extends from Haymarket to Medford and lies between the Orange and Red lines. Three segments were identified for improvements in the corridor: North Station, Lechmere, and Beyond Lechmere.

The 1980 study was undertaken simultaneously with the City of Boston's unveiling of a plan to redevelop the North Station area. Two major components of the city's plans were the construction of a new federal office building and a new multipurpose arena. The Green Line presently rises from subway to elevated structure at North Station. The elevated structure, which is over 70 years old, has been a blight on the area and detrimental to the city's past revitalization efforts. North Station is a gateway to the city and the hub of the North Side's transportation network. The Orange Line serves the commuters to the north; the Green Line serves Cambridge and Somerville; and commuter rail serves the communities farther out to the north and northwest. In addition, many bus routes from the north now terminate nearby at Haymarket Station.

North Station

The City of Boston's redevelopment plans provided a unique opportunity for transportation improvements at North Station.

Initially, the MBTA identified eight alternatives to relocate the Green Line. An alternatives report and a draft Environmental Impact Statement were completed in 1982. Commuter rail improvements at North Station were expected to be a separate project but common to all Green Line alternatives. The following is a brief description of each alternative and the rationale for giving it or not giving it further consideration.

1. Alternative 1—No-Build: Alternative 1 would have maintained the existing Green Line rapid transit service and facilities in the North Station

TABLE 2 LIGHT RAIL OPERATIONS SCHEDULE: PEAK PERIOD

	1988			1990			Post-1990		
	Trips	Consist	Headway (min)	Total Cars	Trips	Consist	Headway (min)	Total Cars	Trips	Consist	Headway (min)	Total Cars

	Trips	Consist	Headway (min)	Total Cars	Trips	Consist	Headway (min)	Total Cars	Trips	Consist	Headway (min)	Total Cars
Boston College (via Commonwealth Ave.)	18	2	5	36	9	2	6	36	12	2	5	42
Cleveland Circle (via Beacon St.)	13	2	6/7	26	9	2	6	30	9	2	6	30
Riverside (via Highland Br.)	14	2	5	37	10	2	6	47	13	2	5	53
	3	3	–	–	–	–	–	–	–	–	–	–
Arborway (PCC) (Forest Hills)	6	1	6	16	10	2	5.6	20	10	2	5.6	20
Arborway (Brigham/Heath)	5	2	–	–								
Blandford Lechmere	–	–	–	–	6	2	10	12	6	2	10	12
Run as directed (RAD)	10	1	–	10	2	1	–	2	2	1	–	2
Totals				125				147				159
Average subway headway (sec)			75				75				62	

area. It would involve no physical modifications to either the elevated or the ground-level station facilities.

2. Alternative 2—At-Grade Relocation: Alternative 2 provided at-grade service between Canal Street and the elevated structure at Science Park Station following the existing alignment or two potential alternative at-grade alignments. This alternative was not carried forward because at-grade transit operations would disrupt vehicular and pedestrian circulation within the North Station district, an area already suffering from vehicular congestion and numerous vehicle-pedestrian conflicts.

3. Alternative 3—Elevated on New Alignment: Alternative 3 provided a new elevated structure between the existing transition section near Canal Street and Science Park Station by way of a new elevated alignment, which would pass between the Boston Garden and the Anclex Building and then run parallel to the elevated Central Artery/Leverett Circle connector ramps to Science Park Station. Alternative 3 was selected for further study because it featured a station location that would facilitate intermodal transfers to commuter rail services and would also serve proposed development in the North Station district. Its alignment was almost totally within public rights-of-way, and its estimated construction cost was about half that of several subway alternatives.

4. Alternative 4—Subway Under Existing Alignment: Alternative 4, which proposed a subway under the existing elevated alignment, was not carried forward for further study. Construction of a subway beneath the existing viaduct, while maintaining present Green Line service above, would present extreme problems related to underpinning and structure security. While technically possible, this construction process would be extremely costly and time consuming.

5. Alternative 5—Subway Under Boston Garden: Alternative 5 provided a below-grade alignment that extended from Haymarket Station, beneath the Boston Garden, and then climbed to meet the elevated Science Park Station. This alternative was further studied and became the preferred alternative.

6. Alternative 6—Subway to Cambridge: Alternative 6 was a subway alignment similar to Alternative 5. Instead of making the transition to the elevated Science Park Station, the alignment continued under the Charles River in a tunnel and ultimately transitioned to Lechmere Station in East Cambridge. This alternative was not studied further due to the dramatically increased investment requirements associated with building a new subsurface river crossing.

7. Alternative 7—Merrimac Street-Lomasney Way Subway: Alternative 7 provided a subway alignment from Haymarket Station via Merrimac Street and Lomasney Way before making its transition to Science Park Station. This alternative was evaluated further because the alignment was totally within

public rights-of-way and was convenient to the (then-proposed) General Services Administration office building. The relocation of the Science Park Station was required by this alternative.

8. Alternative 8—Replacement Bus Service: Alternative 8 eliminated all Green Line service between North Station and Lechmere Station, and made North Station the terminus for the Green Line. Bus service would have replaced the Green Line service to Cambridge. This alternative was rejected because replacement of light rail with bus did not conform to the stated goals of the MBTA or the Northwest Corridor communities of Boston, Cambridge, and Somerville.

Because of the complexities of the project, a preliminary engineering analysis was undertaken as the initial design step and proved to be invaluable. The alternatives were again examined and a detailed engineering analysis was undertaken on the two most promising alternatives: relocating the elevated alignment that ran beside and behind the Boston Garden (Alternative 3); and providing a subway alignment under the Boston Garden (Alternative 5).

An extensive geotechnical program that included a number of test pits was undertaken. A peer review group was formed and contractors were invited to participate in the engineering analysis. The most difficult part of the subway alternative was the tunnel under the Boston Garden, which has to be kept open during construction.

The engineering analysis showed that the supposedly cheaper option, Alternative 3—the relocated elevated structure—would have such impact on an adjacent building that it would cause its taking at a value of $25 million. Nor would the elevated structure afford the simple modal interchange provided by the subway alternative.

The relocation of the Green Line to a new subway alignment will enhance the change of mode at North Station and create a major transportation center. The North Station Transportation Center will serve the MBTA commuter rail, the Green and Orange lines, commuter buses, taxis, pedestrians, and attendees of Boston Garden events. The transportation center will be more than a location where many transportation modes converge; it is being designed to facilitate intermodal transfers, improve existing facilities and transportation services, and increase user comfort. It is being designed with full understanding of the existing surroundings as well as future plans in order to maximize coordination and thereby minimize conflicts among objectives and projects.

The subway alignment runs parallel to the Orange Line with track spacing of 18 ft as far as the north wall of the Boston Garden. There, it swings to the west, simultaneously increasing the track spacing to provide storage facilities under the MBTA commuter rail tracks. Continuing west, it swings to the north and emerges within the median of the proposed widened Lomasney Way to Science Park Station (see Figure 4).

Vertically, the alignment is governed by the existing profile at Haymarket and Science Park stations, the elevations of the Orange Line mezzanine and platform, and by the outfall sewer in Nashua Street. The profiles of inbound and outbound tracks are different within the station and beyond. The outbound track continues from Haymarket portal to Boston Garden nearly level and at the elevation of the mezzanine and then dips. The inbound track dips from the Haymarket portal to meet the elevation of the Orange Line platform. Beyond Boston Garden the profiles meet and continue nearly level to accommodate storage facilities. At Nashua Street, both profiles climb at constant 6.5 percent grade to Science Park Station.

The proposed Green Line station has been designed to serve existing and projected transit ridership. It will not only improve transit service but will also provide efficient connections with other transit modes, including the Orange Line, commuter rail, buses, taxis, and pedestrian routes. The station will have entrances at both ends of its platforms convenient to major pedestrian flow from the Government Center and financial districts to the south and the Boston Garden/commuter rail terminal to the north.

Entrances will be highly visible, clearly marked, and at ground level to promote security and street-level activity. Access to commuter rail will be provided through a pedestrian passageway under Causeway Street. A shared inbound ("super") platform will connect the Green Line directly with the Orange Line (see Figure 5). Direct connections will also be provided to the bus terminal above the Green Line station.

The station will be designed to provide the patron comfort and visual clarity to help them readily find their destinations. The spatial character of the station will accentuate major decision points such as collection areas, critical circulation elements, and the intersections of main paths.

There will be a four-track storage and turnback configuration behind the Boston Garden with storage space for 11 cars (see Figure 6). The turnback area will provide greater flexibility in handling extra or disabled cars. Also, it will serve as the turnback facility for the cars terminating at North Station. Extra cars will be stored in the area for the surge of patrons from Boston Garden events.

As a result of combining the Orange and Green line platforms, an opportunity exists to bring the existing Orange Line station up to current MBTA design criteria. New handicapped access will be provided with an elevator from the north mezzanine to the Orange Line outbound platform. New wall and ceiling finishes and accessories will be coordinated with the new Green Line portion of the station. The existing substandard portions of the platforms will be widened to a minimum of 8 ft and new access from the south end of

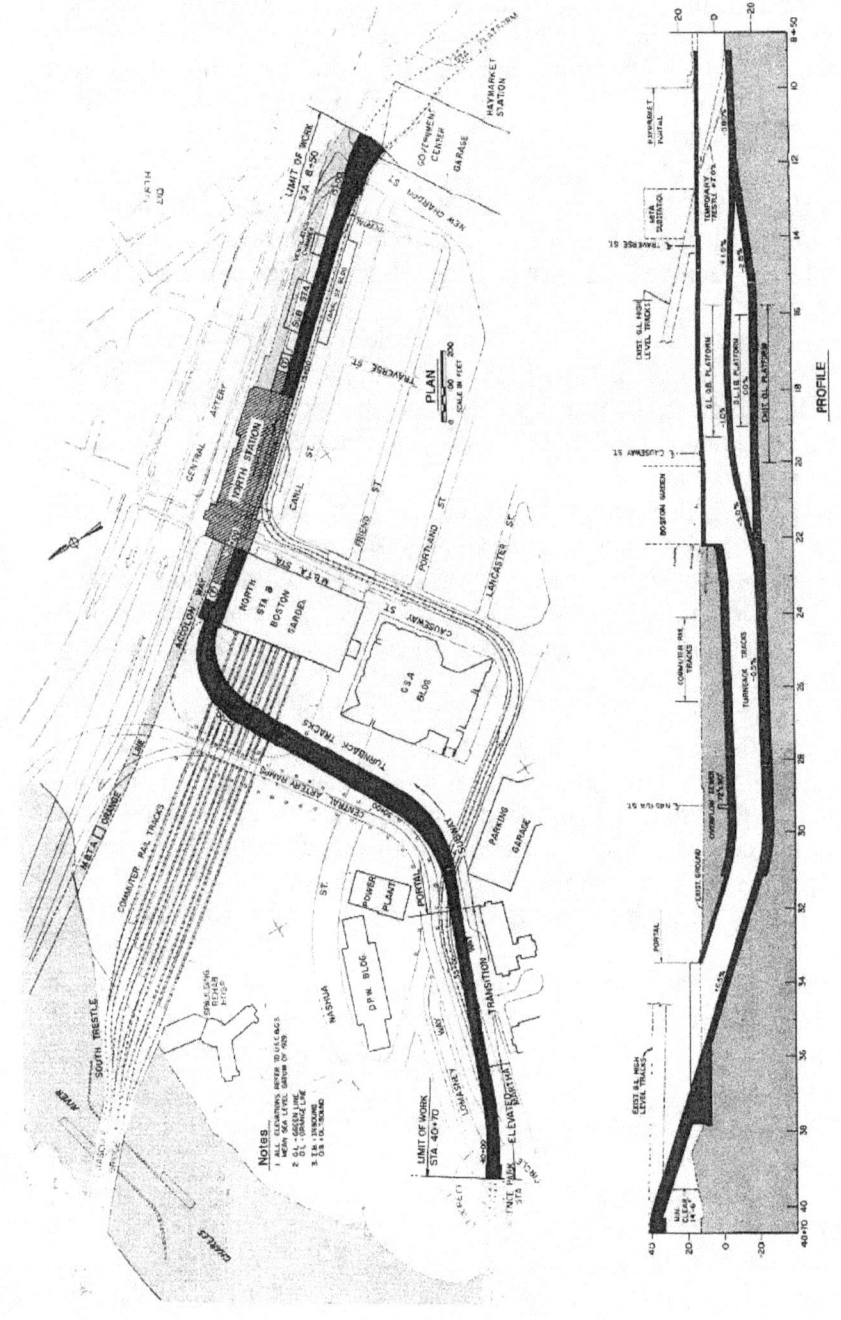

FIGURE 4 North Station plan and profile.

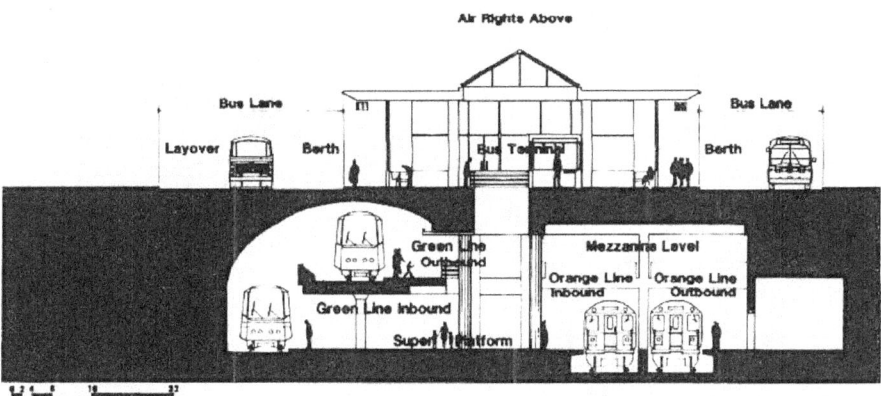

FIGURE 5 New North Station Transportation Center cross section.

the station will be provided via a stair/escalator unit from the new south mezzanine. The roof will be raised to a higher level, allowing natural light from skylights to reach both Orange Line platforms. In addition, all new artificial lighting and graphics will be coordinated with the Green Line portion of the station to provide a uniform, cohesive visual effect within the facility.

The depression of the Central Artery (the major north/south freeway), which presently runs through the city on an elevated structure, will have on and off ramps at Causeway Street across from the new station. The new ramps are ideal for the buses coming from the north and terminating at North Station. A new bus terminal will be constructed at grade above the Green/Orange station to serve bus routes from the north, making the station the best location for the transfer from bus to rail. As previously discussed, the Green/Orange station will have a combined platform for inbound riders and, because both lines provide service to some of the same areas, many transit riders will have the opportunity to take the first train to arrive, whatever color line it runs on.

The development of the station and the bus terminal will create the opportunity to develop the air rights above the transportation center as well. A feasibility study on the potential of air rights that will identify the highest and best use will soon be undertaken; however, preliminary indications are that an office use would be very marketable. The air rights development will provide additional funds for the transit project. In exchange for the air rights, a developer will make a contribution, such as a lease agreement, maintenance, or paying for a portion of the project.

Lechmere Station

Lechmere Station is the northern terminus of the Green Line and is connected to Science Park Station by an arched viaduct across the Charles River. The

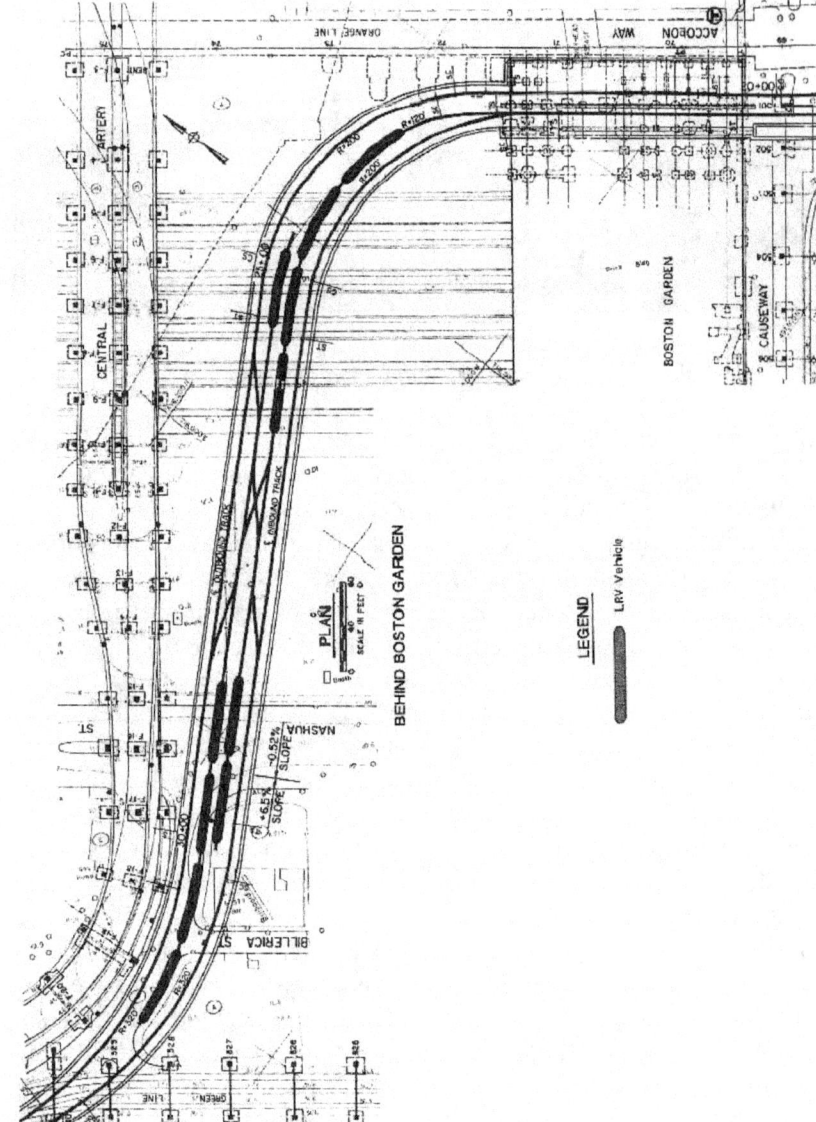

FIGURE 6 North State storage tracks.

arched viaduct was constructed in 1912 and is a historic landmark. The existing Lechmere Station was also constructed in 1912 and has operational deficiencies: lack of storage space, difficult bus movements, and a site that prohibits extension or expansion.

The Lechmere Canal area is undergoing a significant redevelopment. The City of Cambridge, as well as other public and private entities, has invested a great deal of effort and money in the revitalization of this area. The new Lechmere Station is a major component of this effort (see Figure 7). In addition to upgrading Green Line service, the new station will greatly improve the appearance of the area, while encouraging future developments such as the Canal Park project.

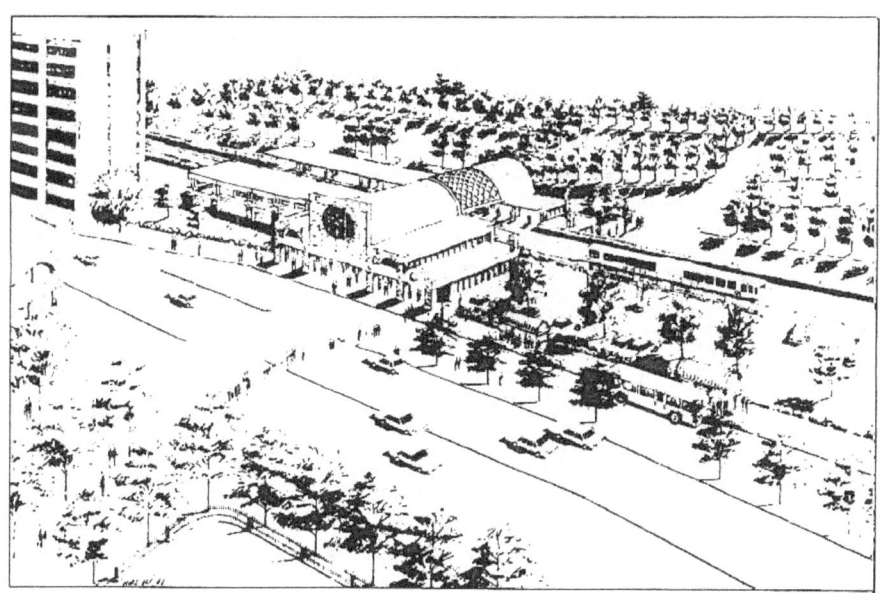

FIGURE 7 Lechmere Station rendering.

The site is primarily occupied now by MBTA parking north of Monsignor O'Brien Highway and across from the existing station. The relocated Green Line track will enter the station area on a viaduct from the east, gradually sloping down to grade level on the west side of the station. The station is located at this transition point on an embankment between elevated and at-grade track.

The relocated station will be highly visible from Monsignor O'Brien Highway and First Street, the major approach routes. The station will form one side of the new Lechmere Square, created by the Lechmere Canal buildings and the development of the existing station site. The eventual removal of the existing station will allow the center of this area to be redeveloped with a combination of open space and a new building.

A major roadway improvement project for Monsignor O'Brien Highway is under way. The relocation of the station will allow further improvements by removing the viaduct from the O'Brien-Cambridge Street intersection and by making other minor improvements possible, such as the upgrading of East Street. Access to the station site will be via East Street. The extension of First Street to O'Brien Highway, a project of interest to the City of Cambridge, would significantly ease traffic flow in the area and help bus and automobile movement to and from the new station.

Pedestrians will cross the highway at-grade at signaled crosswalks. The Cambridge Community Development Department and local East Cambridge groups are interested in a pedestrian bridge that would be fully accessible to handicapped and elderly patrons, and would be located to serve both the East Cambridge community and the Lechmere Canal area.

The station entrance is oriented toward the south and Monsignor O'Brien Highway, the primary approach for pedestrians and motorists. This area also will serve as the drop-off and pick-up area for bus passengers (see Figure 8). A covered platform for five buses will extend from the entrance, parallel to O'Brien Highway. A covered drop-off area will be provided for kiss-and-ride patrons; 300 parking spaces, controlled by one collection booth, also will be provided. A covered walkway will provide a path from the north side parking areas and the industrial development of the North Point area.

The entrance to the station will be through an enclosed brick structure that will contain the pay area, bus waiting, the concession, and vertical circulation. Within this space, access will be provided directly to the inbound rail platform and to a passage under the tracks to the outbound platform. Access to public toilets and the station service areas will be from the passageway under the tracks.

The rail platforms, located on an embankment one level above the entrance, will be reached by way of stairs, ramps, and possibly escalators. Both platforms are to be sheltered, with the track area open.

The building form and the materials to be used in the station are based on those commonly found in the older commercial and public buildings in East Cambridge. Brick columns, walls, and arches, in combination with the concrete viaduct and the glass enclosure and canopies, will emphasize this relationship between the station and the local context.

The construction sequence allows for continuous train service throughout construction. Both tracks can be maintained in operation, servicing the existing station and subsequently the new station, except for a period of 1 to 2 months during the phased rerouting when only one track will be in use.

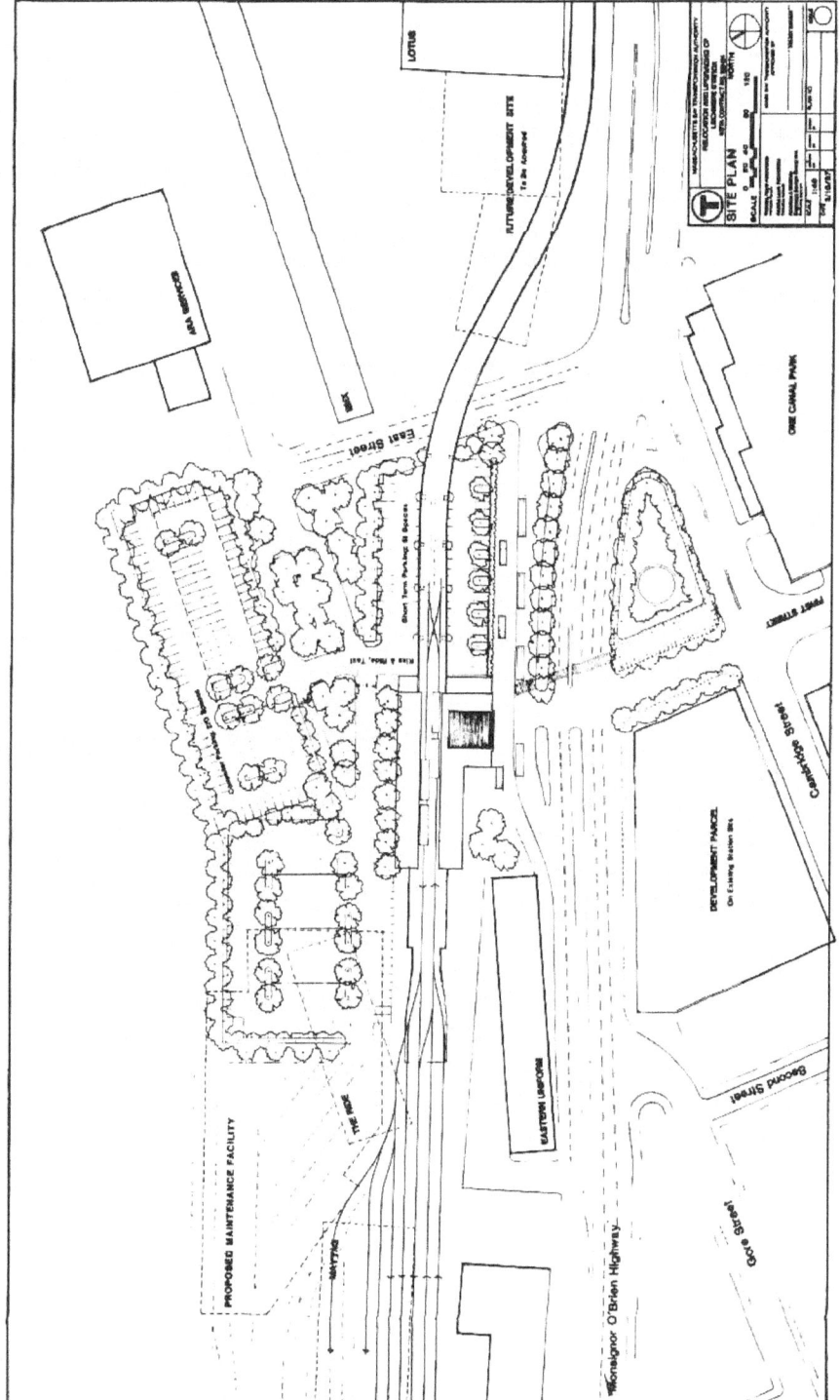

FIGURE 8 Lechmere Station site plan.

The new, relocated Lechmere Station will provide several operational benefits. The new site will be of sufficient size to provide train storage, operators lobby, bus area, maintenance facility, and work train area. In addition, the new station site will be next to the New Hampshire commuter rail right-of-way that may be used for an extension of Green Line service beyond Lechmere into Somerville and Medford.

Initially, a three-level station and LRV storage on a viaduct were studied, but emphasis on the related maintenance facility favored the current two-level embankment station. With the current station design, the related LRV storage can occur at grade rather than on viaduct, the connection between the rail line and buses is improved, the maintenance facility can be closer to the station, and the overall cost is significantly lower.

Lechmere Maintenance Facility

The Green Line is one of the largest light rail operations in North America, with four branches merging from the west into the Central Subway to downtown Boston and then north to a terminus at Lechmere. But vehicle maintenance deficiencies exist in the present system. All the LRV maintenance facilities are located at the western terminus points at Riverside and Reservoir with a running repair shop at Boston College. This arrangement requires all disabled cars running from the Central Subway to be moved a significant distance for repairs.

The existing Green Line facilities cannot provide the levels of maintenance and storage needed to support a larger fleet and expanded service. Nor can they be economically enlarged to satisfy increased requirements. A new LRV maintenance facility at Lechmere would be ideally located near downtown and the Central Subway. The Lechmere site is directly accessible to all branches and would produce a significant savings in car miles. It would also greatly improve the flow of disabled cars to be repaired, especially for failures occurring inbound in the Central Subway. In addition, the new Lechmere facility will provide a secondary benefit to Green Line operations by reducing the backlog of cars waiting to be repaired at the already over-taxed Riverside and Reservoir facilities.

Maintenance

Maintenance functions can generally be divided into the following areas:

- Running repairs,
- Periodic inspections (performed every 30 days),

- Annual inspections, and
- Heavy repairs (which include numerous categories and take more than one day to perform).

A recent review of shop records for two time periods showed an average of 50 cars out of service. Of this total, 32.5 or two-thirds were projected to be out of service 1 day or less, 20 percent for 2 to 5 days, and 14 percent for 6 or more days. It is estimated that there are approximately 40 maintenance actions per day, the bulk of which are running repairs.

A statistical summary of the three principal maintenance facilities on the Green Line—Riverside, Reservoir, and Boston College (Lake Street)—is shown in Table 3.

TABLE 3 GREEN LINE MAINTENANCE AND STORAGE FACILITIES

Carhouse	Running Repair Spots	Heavy Repair Spots	Yard Storage Capacity
Lechmere	—	—	18
Riverside	12	20	72
Reservoir	12	—	62
Boston College	2	—	21
Total	26	20	173

Storage

To determine the requirements for storage at Lechmere, several car-flow plans were developed. Essentially, it was determined that 40 to 44 cars were to be left at Lechmere during midday storage. The car-flow plans require that some trains be operated on different branches during a run. Although this is often done on an unscheduled basis, it is a change from current scheduling practice. This change will prevent any scheduled headway gaps or increases in car miles.

Lechmere Yard Storage Requirements

In addition to the midday storage, space has to be provided for storage of spare cars and for shop support. The 1990 service plan calls for the number of spare cars to be about a third of those operating. It would be operationally unwise to assume that all spare cars would be kept at Reservoir or Riverside. Therefore, some spare car space should be provided at Lechmere. The number of spaces required for shop support should permit resetting the shop on a given day.

The planned overnight storage at Lechmere, exclusive of the spare cars and the shop support, is as follows:

Storage	No. of Cars
Heath Street	20
Blandford Street	12
Run as directed (RAD)	2
Total	34

If one-third of these cars are designated as spares, about 10 spaces would be required to store them. Therefore, the estimated 1990 storage requirements for Lechmere is as follows:

Storage	No. of Cars
Midday	45
Spare	10
Shop support	15
Total	70

Because midday storage requirements exceed the overnight storage requirements, the space may be used to begin morning start-up service from Lechmere for other lines, too.

Shop Requirements

The ultimate shop requirements for the Green Line depend upon a number of factors. For example, by the year 2000, the Boeing LRVs will be over 25 years old and candidates for replacement. Thus, the composition of the fleet could be significantly different than it is today. Given this uncertainty, the analysis provides general guidelines for the shop requirements with post-1990 service levels.

Assumptions

The following assumptions were used for the analysis:

- The fleet will consist of 250 cars with 200 required for service. This results in an improved availability ratio of 80 percent.
- System car miles would increase in the same ratio as the increase in peak period car requirements. Thus post-1990 car miles will increase by a ratio of 1.33 to 8,342,666 mi.

- Mean distance between failures will approximately double to 3,000 mi.
- Approximately 50 percent of the failures will be sent to Lechmere compared with 40 percent in 1990. The increase is the result of new extensions for which Lechmere will be most accessible.

A summary of the storage and shop requirements at Lechmere based on a preliminary analysis is contained in Table 4.

TABLE 4 MAINTENANCE AND STORAGE REQUIREMENTS AT LECHMERE

Category	Existing	1990	Post-1990
Assumptions			
Active fleet	175	225	250
Peak cars required	105	150	200
Mean distance between failure (mi)	1,300	2,250	3,000
Car miles (thousands)	4,526	6,257	8,343
Maintenance incidents per day (system)	40	37	37
Maintenance incidents—Lechmere	0	15	19
Results			
Storage—Lechmere (cars)	18	70	100
Running repair spots—Lechmere	—	10	13–14
Heavy repair spots—Lechmere	—	15	15
Total repair spots—Lechmere	—	25	28–29

Beyond Lechmere

Travel in the corridor beyond Lechmere to Somerville is strongly oriented towards downtown Boston and neighboring urban centers. Analysis of origin-destination studies reveals that about a quarter of a million trips begin or end in the study area on a typical weekday. While 16 percent of these trips occur entirely within the study area, about 25 percent of the trips are oriented towards downtown Boston and Cambridge. In particular, journey-to-work trips show a strong orientation towards downtown Boston.

Transit accounts for 70 percent of the study area trips made to downtown Boston. An analysis of the demographic profile reveals some of the reasons for this high level of transit dependency and usage. The area has a high population density, a high percentage of elderly and low- to moderate-income residents, and a low level of automobile ownership—all indicators of transit dependency. Given such a high rate of public transit usage, transit system improvements (excluding new ridership from transit-induced new developments) are more likely to provide better service for existing riders than to attract new riders from an untapped transit market.

The corridor is served by an extensive system of buses, which primarily feed Lechmere Station. Ridership statistics indicate that a high proportion of trips originating in the corridor have destinations within it or in the North Station area of downtown Boston. These trips will not be well served by the Orange and Red lines because these heavy rail facilities are too distant and because of the inconvenience caused by the multiple intermodal transfers required to reach them via local bus.

An evaluation report on the alternatives beyond Lechmere was completed in 1984. The report evaluated a number of transit alternatives for the beyond-Lechmere corridor, including light rail, bus, busway, and combination light rail and busway. Most promising of the alternatives is an extension of the Green Line along the New Hampshire Main Line commuter rail route. The New Hampshire Main Line runs through the middle of the study corridor and is of sufficient width to accommodate both commuter rail and the Green Line.

The Green Line extension would be approximately 3.5 mi long and terminate in the vicinity of Tufts University. Although this alternative would not attract a large number of new riders because the area is already heavily dependent on transit, it would provide passengers with a one-seat ride to downtown Boston. One of the operational goals of an extension of the Green Line beyond Lechmere is the reduction of bus miles that would result.

An extension of Green Line service beyond Lechmere can be easily accomplished due to the availability of a portion of the New Hampshire Main Line right-of-way, which is depressed, and the flexibility that comes with light rail. The project can be constructed in segments to meet available funding. Simple platforms with crossovers can serve as temporary stations. No major parking structures or expensive stations will be required for the extension.

CONCLUSION

As we approach the 21st century, the need for mass transit becomes even more demanding. Although recent improvements to the heavy rail lines have increased their capacity and efficiency, Boston's oldest system, the Green Line, must also be improved. New, relocated facilities at the Lechmere and North stations are the first improvements. The new North Station will provide riders with improved transfer capabilities and operations with much needed storage and turnaround facilities for the LRVs. The relocated Lechmere Station will provide the opportunity to develop an LRV maintenance facility for the growing fleet and to extend service beyond Lechmere into Somerville. After a century of service, Boston's light rail is still looking to the future.

Rail Start-Ups

Having the Right People in the Right Place at the Right Time

Peter R. Bishop

Staff plans and practices are vital to the success of any light rail operation. In Buffalo, the Niagara Frontier Transportation Authority's Metro Rail system began its efforts to get the right people in the right place at the right time in 1981 with a nationwide search for a rail operations leader with a background in research and development. With this superintendent aboard two years before the system began revenue service start-up tasks such as developing a rule book and standard operating procedures began. Management personnel were recruited next and sent to the Port Authority Transit Corporation's facilities in New Jersey to learn from an operating light rail system. Filling the rest of Metro Rail's positions then began. Screening for nonunion employees was extensive and systematic. Union employees recruited from Metro's bus operations, however, could only be ranked by seniority. Training became the next consideration and was at times complicated by the fact that, although equipment had been delivered, not all of it was operational when expected. The success of the recruitment and training process shows up in Metro Rail's low turnover rate.

THE NIAGARA FRONTIER TRANSPORTATION AUTHORITY (NFTA) was created by an act of the New York State Legislature in 1967. The NFTA, a public-benefit corporation owned by the citizens of New York, was assigned responsibility for developing air, water, and surface transportation in Erie and Niagara counties. The authority was given the further mission of formulating and putting into effect a unified mass transportation policy for the Niagara Frontier.

Niagara Frontier Transportation Authority, 93 Oak Street, Buffalo, N.Y. 14203.

Between 1967 and 1970 the NFTA conducted extensive planning studies leading to the production of the Transit Development Program. This program, which was officially approved by the legislature in 1971, encompassed three major elements: the establishment of a regional bus transit network, the construction of a Metropolitan Transportation Center in downtown Buffalo, and the design and construction of a rail transit system between Buffalo's waterfront and the suburban community of Amherst.

During the early 1970s the design of such a rail transit system was planned as a heavy rail line operating through the principal urban corridor, with both subway and aerial structures. After intense local review, the route ultimately evolved as a combination light and heavy rail system with mall operation through the central business district (CBD) and in tunnel elsewhere. The substitution of tunnel for aerial structures, occasioned by community opposition to the latter, substantially increased the cost of the project. As a result, to remain within fiscal limits the length of the line had to be reduced. Instead of a 10-mi heavy rail line, the project was scaled down to a shorter light rail route located completely within the city limits.

Construction of the Metro Rail line began in 1979. The project created many hundreds of badly needed jobs and much of the total cost of $530 million was spent in the western New York area. As a public works effort, Metro Rail surpassed in size even the famous hydropower installations at Niagara Falls.

Metro Rail opened in stages. Operation through the downtown mall, itself under construction, began on October 9, 1984. On May 18, 1985, trains began to operate underground as well for a total distance of 5 mi from the downtown terminal. On November 10, 1986, the entire route from Memorial Auditorium to the south campus of the State University of New York at Buffalo was opened to the public.

The current Metro Rail line consists of 6.2 dual-tracked route miles, 27 double-ended cars, 8 architecturally distinctive subway stations, and 6 stations located along the world's largest pedestrian mall, Buffalo Place. By early 1987 this modest-sized rail operation was carrying 30,000 daily riders. It currently operates weekdays and Saturdays until midnight with limited Sunday service. During peak periods the trains run every 6 min. Between the morning and afternoon peak periods, the trains operate on 10-min headways, while in the base periods the headway is lengthened to 20 min.

In 1981 Metro Rail was faced with its initial application of the "right people in the right place at the right time" rule. A company philosophy for preliminary staffing necessitated a nationwide personnel search for a rail operations leader with a background in research and development. Revenue service was still 3 years away but there was a pressing need to put in place as many operational facets as possible. The very heart of the rail transportation department began beating in 1982 when Anthony Schill came on board as superintendent. Through his efforts over the next 2 years, a rule book was developed, standard operating procedures were written, administrative and operational forms were designed, and a myriad of other start-up tasks were shouldered by the superintendent as the slow transition from construction to actual revenue operations got underway.

This transition developed its own set of prioritized problems. It also pointed up the importance of putting together a staff that could call on others in the industry to seek out knowledge and experience. Where possible, Metro Rail personnel wanted to avoid the problems cited by their contemporaries.

Supplemental staffing began in earnest in 1984 when employees from bus transportation, who would ultimately be responsible for various areas such as training, supervision, and operations, were reassigned to rail transportation. The superintendent's rail background was thus complemented by his staff's company background and experiences.

Metro decided that rail should be an entity separate from the established bus operations. This allowed the superintendent greater flexibility in performing his duties. However, it also developed a division within operations that is currently being evaluated. Initially such a division was appropriate, owing to the differences between the operating techniques of bus and rail. It also served as an enticement for recruitment within the rank and file. But now that a safe and efficient rail transportation has been operating for 4 years, the transportation division is taking steps to integrate rail with bus in all appropriate areas.

Once the initial candidates for transfer to rail management had been identified and approved, it was necessary to establish a training program. Rail familiarization and indoctrination for these management employees were accomplished during a 4-week tour of the Port Authority Transit Corporation (PATCO) facilities in Lindenwald, New Jersey. Basic concepts of training, operation of rail vehicles, record keeping, and ancillary functions such as maintenance, revenue, public relations, and control tower operations were viewed. More important, friendships were fostered that proved invaluable in the future. It was Buffalo's intention not to mirror PATCO's operation but to witness a successful operation and define those general concepts that could be applied to the new system.

Once back in Buffalo, management's attention was redirected to the immediate problems of accepting equipment and making it operational. Development of a training program dealing with the new equipment was an obvious requirement that resulted in its own unique set of problems. Personnel positions, both union and salaried, had to be structured. A system of recruitment for those positions from the rank and file was set up. All of this was accomplished against a timetable that kept slipping.

Operating personnel at Metro Rail are classified as either union hourly employees or nonunion salaried employees. The rail operations positions staffed by union employees are those of train operators, ticket inspectors, and station clerks. The rail operations positions staffed by salaried employees are those of train controllers and rail supervisors.

The train operator is responsible for proper and authorized operation of a rail vehicle in conformance with a published schedule of movements as well as other duties. The ticket inspector is responsible for passenger compliance with all published regulations regarding fare payment as well as other duties. The station clerk, an administrative position, is responsible for the distribution of work to train operators and all other clerical duties associated with the conduct of daily business at the station. The train operators and station clerks report to the district manager and are governed by a 3-year labor agreement between Metro Rail and the Amalgamated Transit Union Local #1342. (Union members are also covered by state legislation that prohibits public employees from engaging in labor strikes.)

The train controller is responsible for the operation of the sophisticated electronic and computer-based equipment that governs the movement of trains, controls traction power, and the ventilation in the tunnel. Rail supervisors are responsible for monitoring the train operators' proper attention to all rules, orders, and procedures that affect train movement. The train controllers and rail supervisors report to the operations control center manager (now retitled the assistant superintendent, rail transportation).

The balance of this presentation deals primarily with train controllers and train operators. The original operations equipment arrangements called for three train control consoles and four station control consoles. Based on this configuration, the initial manpower staffing levels would have required 10 controllers. The original hours of operations would have required four supervisors. Today's operations use fewer controllers, and more supervisors, and transit police personnel have been added.

During the early stages of the transition from construction to revenue service, a general mandate to cut costs triggered a review and subsequent reduction of staff and a shifting of responsibilities. The original complement of 10 controllers was cut to five. One console each was cut from the train control and the station control areas. The responsibilities for station operations were reassigned to the transit police, however, with the associated overhead costs retained by the rail transportation department.

The first two train controllers were recruited from outside the company. Individuals were solicited with previous rail operations experience, particularly those with control tower backgrounds. This was consistent with Metro Rail's philosophy of employing some "rail" people who could act as trainers. These two individuals, one from the Lehigh Railroad and the other from the Southeastern Pennsylvania Transit Authority (SEPTA), were joined by three trainees from within the ranks of Metro. The in-house people had a variety of backgrounds. They included a bus operator, a schedule department clerk, and a transportation assistant from the bus operations department. This original nucleus of five controllers has now been expanded to seven due to the additional responsibilities associated with increased hours of operation.

All in-house Metro candidates for controller positions responded to a job posting displayed throughout the company's premises. The response was overwhelming to the point that a screening process had to be developed that could reduce the number of applicants to a manageable level. It should be noted that this position was and remains a nonunion position. The basic qualification sought by the recruitment process was trainability. This was determined by weighing an applicant's company seniority, experience, and education. Work records of all applicants were reviewed for disciplinary actions. Interviews were conducted and innate mathematics, vocabulary, and comprehension testing was completed. Final recommendations were made to the internal personnel selection committee for approval. These steps led to the appointment of trainees who proved very trainable. During 4 years of operation, the turnover rate has been extremely low. Only one of the original outside controllers has left, returning to his former employer. This outstanding retention is directly attributable to the early screening efforts.

To refurbish a cliché, equipment waits for no trainee. Metro was optimistic to expect that, as personnel became available, the equipment they would work with would also become available. In reality, the equipment was shipped in time, but was not operational in time. The silver lining in this cloud was that the controllers not only had to learn how the systems were supposed to work, but also how they actually did work and what had to be done to make those two concepts compatible. This required considerable mental agility and the development of crisis management techniques. Such a grounding in basic operations is still bearing fruit today. On-time performance associated with terminal departures exceeds 99 percent efficiency.

The hardware and software necessary for our operation were termed "the leading edge of technology." At the time of installation that was probably true and as such the chief benefits Metro Rail derived from the original two controllers were not only having them act as trainers but also having them keep the system operational under very rudimentary conditions. While they were doing that, the balance of the controllers were learning basic railroad operations along with the new technologies.

Finally it all came together and regular routines were developed to cover Metro Rail's commitment to the riding public. During peak periods, which are from 7:00 a.m. to 8:30 a.m. and from 2:30 p.m. to 5:30 p.m., two controllers are working—one for the surface activity and the other for the subsurface activity. The balance of the day requires one controller to monitor the entire system. The controllers were given an opportunity to direct their own destiny when they offered suggestions on scheduling their work. Management listened, the controllers contributed, and thereby a measure of stress was removed from the control center. Currently 8-hour work assignments, which last 4 weeks, are selected by the controllers every 8 weeks. The order of selection is governed by a sliding seniority list. The only provision is that each controller must meet a quota of overnight assignments during a 6-month period.

All controllers are required to be certified every year, not only in their specific disciplines but also as train operators. Periodic train operation in revenue service is encouraged. This gives the controllers a better appreciation for the actual operating environment as seen through the eyes of a train operator. It also provided additional train operators when vacations, illness, and line practice training combined to produce a shortage of regular operators during summer 1987. No runs were cut.

Train controller recertification takes place annually. It is conducted by the assistant superintendent, who is responsible for the preparation of recertification criteria. Testing of rules, procedures, techniques, and experiences is required. This usually prompts the controllers to review annually those items that are a little vague before the testing begins.

Due to the expansion of operating hours Metro Rail has increased the original complement of five controllers. Using the established screening methods, we have added two more controllers. One was a bus operator and former yardmaster on the South Buffalo Railroad. The other was schedule designer from the service planning department. They have blended well with their coworkers. Their training was the responsibility of their fellow controllers, who developed an early relationship with them, nurtured through on-the-job training. The trainer-controllers were also able to relate real life experiences that helped to give a realistic perspective on the world of computers and electronics.

When a posting for 20 train operators was displayed, 134 employees responded and were placed on a trainee list. Their selection from the ranks of bus operators was less sophisticated because our attempts to use a screening method similar to that used for the train controllers were thwarted by the union. The only acceptable criterion was to rank the train operator applicants by seniority.

The train operator training program conducted by Metro's training department is 4 weeks long. The initial sessions deal with a concentrated review of the rule book, standard operating procedures, and all current orders and notices. This review features various written tests. Classroom sessions are supplemented with vehicle operation sessions. The second and third weeks primarily provide "seat time" for the trainee operators. They are accompanied by a line instructor on trains not in service. Not only do the trainees gain familiarity with the rail cars but also with distance perception for speed and braking purposes, station announcements, troubleshooting, and yard techniques. The last week of training consists of actual revenue operation with another regular operator present and the final comprehensive tests on signals and the rule book. The trainees are given only two opportunities to pass these final tests. If they don't succeed, they are washed out of training.

The collective bargaining agreement in place during the period in which Metro Rail was recruiting train operators made no provisions for rail transportation. A separate memorandum of agreement was required. One of the original stipulations was a commitment from the permanent train operators of at least 3 years of service. Now that we are approaching the end of that time limit for some operators, Metro Rail has had to reevaluate its personnel replenishment program. The original concept was to train as many permanent train operators for assignment as dictated by the scheduled service. However, service needs expanded faster than manpower levels. It was thought that temporary train operators could fill these short-term voids and also become a reservoir for future long-term needs. Therefore a request for temporary train operators was posted throughout the company. This resulted in a trainee list separate from the permanent operators' trainee list.

The agreement also stated that temporary train operators could only refuse a permanent train operator position twice before being banned from the rail operations. This created a problem for the company because the trainees apparently were interested only in an exposure to rail operations. They did not want to relinquish their seniority position at a bus station for lesser privileges, relative to run selection, at the rail station. This lack of commitment from the temporary train operator trainees meant more trainees had to be processed than there were open positions.

Many other items were spelled out in this memorandum of agreement but experience showed that its best feature was its expiration shortly after the commencement of full-scale operations. All segments of union activities that relate to rail transportation are now included in the current collective bargaining agreement. This made contract negotiations a little lengthier, but the result was well worth the effort. To this day Metro Rail is still finding nuisances that are not covered, but special provisions for these can be made. Actual operation is a wonderful test ground for such things as relief points, turn-in times, and report times.

Every train operator must be recertified annually by the training department. Such things as additions or deletions to the rule book, the standard operating procedures, and the operations orders are reviewed and tested. Actual train operation is monitored and bad habits that have crept into the train operators' techniques are corrected. System safety and emergency procedures are emphasized and troubleshooting methods are discussed. Retrofittings to the rail cars are explained and demonstrated. All questions are answered and test results are documented. With this program in place we have minimized our accident record and developed better harmony between operators and management.

Our work with the operators also includes informal "rap" sessions. These meetings between union members and management are usually held on a Sunday morning and attendance is voluntary and not compensated. Gripes are aired, suggestions for improvements are offered, and problems are resolved before they reach the grievance stage. These meetings definitely contribute to the sense of family at Metro Rail.

Everyone in Metro Rail understands that their efforts to make the system a success are well directed. Metro Rail is a tremendous catalyst in Buffalo. With the support this earns from the public, it is easy to understand our boasting that we have the right people in the right place at the right time.

Lessons Learned from New LRT Start-Ups

The Portland Experience

RICHARD L. GERHART

The first light rail line in Portland, Oregon, began revenue service on September 5, 1986, after more than a decade of planning, engineering, and construction. The project was known as the Banfield Light Rail Project, recognizing the combined scope of Banfield Freeway (I-84) improvements and light rail construction. The combined $319-million project, jointly managed by the Oregon Department of Transportation and the Tri-County Metropolitan Transportation District of Oregon (Tri-Met), was the largest single public works project in the state's history. The overall project was delivered on schedule and within budget. The successful start-up of the 15.1-mi Portland-to-Gresham line was accomplished by stressing teamwork throughout all phases of the project. The transition from engineering staff to operating personnel was structured to maximize coordination. The establishment of an operations core start-up team provided the organizational framework necessary to develop a rail operations plan and complementary start-up activities schedule. First-year ridership exceeded prerevenue service estimates, and operating costs were below budget. This success reflects the importance Tri-Met assigned to learning as much as possible from properties with experience in light rail operations, and to including all areas of Tri-Met's organization in the development and activation of the start-up plan.

Tri-County Metropolitan Transportation District of Oregon, 4012 S.E. 17th Avenue, Portland, Oreg. 97202.

THE TRI-COUNTY METROPOLITAN TRANSPORTATION District of Oregon (Tri-Met) is the public transportation agency in the Portland region. Tri-Met serves a 725-mi^2 service area in three Oregon counties (Multnomah, Washington, and Clackamas). The service area population is slightly less than 1 million. Tri-Met has a fleet of 550 buses, of which 87 are articulated, and 26 articulated light rail vehicles (LRVs). The LRVs operate on a 15.1-mi rail line between downtown Portland and the City of Gresham, located in east Multnomah County.

Tri-Met was created by the Oregon legislature in 1969 to acquire the assets of the privately owned systems then providing transit service in Portland and its suburbs. It has a seven-member board of directors appointed by the governor. In addition to farebox revenues, Tri-Met is financially supported by a payroll tax levied at the rate of 0.6 percent on all employers' payrolls and self-employed persons in its service area. (There is no sales tax in Oregon.)

The Tri-Met system transports approximately 120,000 originating ("revenue" or "linked") passengers each weekday. About 60 percent of these trips are generated in the more densely populated area of the City of Portland; the remainder is almost all suburban ridership. During the peak hour 411 buses and 22 LRVs (11 two-car trains) are in service. More than 40 percent of the peak hour work trips to the Portland central business district (CBD) are made on Tri-Met.

Portland's light rail transit (LRT) system is the result of a freeway construction controversy that occurred in the mid-1970s. As part of the federal Interstate highway network, the Oregon Department of Transportation (ODOT) had proposed construction of the Mt. Hood Freeway. The name of the proposed freeway was somewhat misleading in that this was actually to be an urban freeway through southeast Portland. The political debate triggered by the freeway proposal resulted in a regional decision to withdraw the freeway proposal and to transfer the funding to a transit-oriented transportation solution. Eventually this produced a $105-million upgrading of a segment of the existing Banfield Freeway (I-84), and the $214-million 15.1-mi LRT system. The LRT system opened on September 5, 1986. It was named MAX, short for Metropolitan Area Express.

MAX has been recognized as a major success from opening day, with average weekday ridership at 20,000 boarding rides (versus a first-year projection of 17,000) and operating and maintenance costs 22 percent below budget for fiscal year 1986–1987. Much of the immediate success of MAX can be attributed to the positive momentum generated by delivering the largest public works project in Oregon's history (the $319-million combined light rail and Banfield freeway widening project) on time and on budget, and by holding an opening weekend celebration, featuring free rides on MAX, that attracted over 150,000 people.

The successful start-up of MAX was really the culmination of more than a decade of planning and coordination. In retracing the history of the project, it becomes obvious that significant lessons were learned in all functional areas of the project (financing, preliminary engineering, construction, etc.). The primary focus of this paper is on the last 2 years before the start of revenue service in September 1986. This 2-year time frame provides an opportunity to critique the most intensive period of rail start-up activity.

PHYSICAL DESCRIPTION AND OPERATING CHARACTERISTICS

MAX extends 15.1 mi in a generally east-west direction between downtown Portland and Gresham. In the Portland city center the line terminates in a three-track offstreet loop just west of 11th Avenue between Morrison and Yamhill streets. Downtown operation on restricted lanes of city streets is on Morrison (westbound) and Yamhill (eastbound) between the 11th Avenue terminus and First Avenue, and on First Avenue between Yamhill and the approach to the Steel Bridge (1).

The line crosses the Willamette River on the Steel Bridge, a double-deck lift span, sharing roadway space with vehicular traffic. On the east side of the river the route stretches about 0.7 mi on a restricted portion of Holladay Street to the start of a completely grade-separated 4.9-mi section between the rights-of-way of the Banfield Freeway (I-84) and the Union Pacific Railroad. This section is between Lloyd Center and Gateway stations.

At Gateway the route crosses over the Banfield Freeway, running then in a north-south direction, adjacent to the I-205 connector freeway, for 0.6 mi between Gateway and Burnside Street. The line then resumes its generally east-west alignment in the median strip of East Burnside Street for 5.3 mi between I-205 and 199th Avenue. From this point to the eastern terminus at Cleveland and Eighth in Gresham, the line runs a distance of 2.1 mi on the former right-of-way of the Portland Traction Company.

Traction power at nominal 750 volts dc is transmitted to cars through simple trolley wire (in the downtown area) or catenary (in the outlying sections). Power is supplied by 14 mainline substations plus one at the Ruby Junction Operations Facility. These unmanned substations use transformer-rectifier units to convert 12,000-volt ac power, provided by Pacific Power and Light Company and Portland General Electric Company, to the 750 volts dc required for operation.

On most of the route the line is double-tracked, providing for one-way travel on each track under normal operating conditions. There are two major exceptions. The easternmost segment of the line, the 2.1 mi between Ruby Junction and Gresham Terminal, is a single-track section with a passing track at Gresham City Hall and a second track at the Gresham Terminus. The line also operates on a single track in the downtown area in a loop, using Morrison Street (westbound) and Yamhill Street (eastbound) between First and 11th avenues.

Track gauge is railroad standard, 4 ft 8½ in. (1435 mm). Between the Gresham Terminal and Lloyd Center, 115-lb heat-treated RE rail is laid on wood ties. Between Lloyd Center and the downtown terminus, girder rail is installed in a latex plastic material that holds rails in position, dampens vibration, and mitigates electrical current leakage.

Crossovers between inbound and outbound tracks are provided at intervals to permit operation in both directions on a single track during trackway repairs or service disruptions. Extra track space is available for emergency or special storage of cars at both terminals and at Coliseum, Hollywood, and Gateway stations.

Rail operation is protected by automatic block signal (ABS) systems in two high-speed sections, one between Lloyd Center and Gateway Station, alongside the Banfield Freeway, and the other between Ruby Junction and the Gresham terminus, the single-track section on the former Portland Traction Company right-of-way. In these sections trains are kept separated by operators' visual observations of wayside signals. Trains are stopped in the event of failure to observe signals, employing automatic train stop (ATS) protection. There is also a short signalized section governing the operation over the Steel Bridge with ATS protection.

In the sections of the route along East Burnside Street and Holladay Street, the line is not signalized per se, but operators are governed by street traffic signal indicators at the numerous intersections. LRVs preempt these signals as they approach, which halts cross traffic and permits the LRVs to proceed through the intersections without stopping. Special bar-type signals, located both in advance of and at each intersection, indicate to the operator whether street traffic signals have been preempted, providing sufficient time for stopping in the event of failure to preempt.

In the downtown area, LRVs are governed by traffic signal indicators, and there are no arrangements for preemptions. The only special rail signal is located at the entrance to the 11th Avenue loop; that signal indicates the status of the switches governing access to the three tracks within the terminus loop.

In downtown Portland, LRVs are scheduled to operate at low speeds (15 mph maximum), controlled by street traffic signals. The traffic signals compose the only crossing protection. On Holladay Street and on East Burnside Street, traffic signals control LRV, pedestrian, and automobile traffic flow at crossings. As noted previously, signals are preempted by approaching trains. In the section along the Banfield Freeway, there are no at-grade crossings. At

199th Avenue and at intersections east of it, grade crossings are protected by gates activated by the arrival and passage of trains. There are 10 locations at which gates are installed.

There are 22 station stops in each direction on the line, requiring 38 station platforms or sidewalk loading locations. (There are several island-style platforms that serve both directions of travel.) Stations are of simple design, generally consisting of concrete slab (or sidewalk in the city), a row of shelters, ticket vending and validating machines, information displays, and a hydraulic lift to raise wheelchair passengers from the platform to the level of the LRV floor. Four stations have park-and-ride facilities, providing a total capacity of about 1,600 parking spaces.

Fares are not collected on trains. Ticket vending machines at each station provide tickets for passengers without transfers or monthly passes. Discounted multiple-ride tickets are available in lots of 10; these tickets are individually validated by passengers on the platform before they board the train.

The fare structure is the same for both Tri-Met buses and MAX; fares are transferable between bus and MAX. Fare inspectors check payment receipts or passes to enforce correct fare payment, and issue citations with court authority to anyone without valid proof of fare payment.

The center of operations for MAX is the operations facility located in a four-story building close to the mainline at Ruby Junction (199th Avenue). The building houses the rail operating staff, the control center for rail, and the report facility for train operators. The facility is also the center of maintenance activities for right-of-way track, signals, and electrical systems, as well as for the LRVs.

Yard tracks surround the operating facility, providing storage space for cars not in service and permitting movement of LRVs to and from the mainline and through the shop and carwasher. The facility also has a storeroom for the spare parts and units required for the maintenance of facilities and equipment.

The LRV passenger fleet consists of 26 double-ended, six-axle articulated cars, with four double doors per side. The manufacturer is Bombardier, Inc., employing a design by BN of Belgium. Car specifications are as follows:

Length, 88 ft;
Width, 8 ft 8 in.;
Height, 12 ft 5 in.;
Floor height, 3 ft 2 in.;
Empty weight, 45 tons;
Seats, 76;
Capacity (seated plus standing passengers), 166 (design load);
Wheelchair spaces, 2;

Maximum speed, 55 mi/hr;
Minimum radius curve, 82 ft;
Brakes, dynamic, disc, and magnetic.

MAX is operated as a regional urban and suburban trunk route. Service is provided between approximately 5 a.m. and 1 a.m. 7 days a week. Service frequencies and train lengths (one- or two-car consists) are designed to provide seats for all passengers in any normal 30-min period in the off-peak period on weekdays and all day on Saturdays, Sundays, and holidays. During weekday peak periods, 7 a.m. to 9 a.m. and 4 p.m. to 6 p.m., service is designed for 30-min car loadings not to exceed 166 passengers per car.

On the basis of the above design standards, weekday MAX trains are scheduled every 7 min during peak periods, every 15 min during off-peak periods and until 10:30 p.m., and then every 30 min until 1 a.m. The peak vehicle requirement is 22 LRVs, deployed as 11 two-car consists. (Consists are limited to a maximum of two LRVs because downtown Portland city blocks are only 200 ft long.) On Saturdays, MAX trains run every 15 min until 10:30 p.m., and then half-hourly until 1 a.m. For Sunday or holiday service, MAX trains are scheduled every 15 min until 7:30 p.m., and then half-hourly until 1 a.m. Single-car trains are typically deployed during weekends, but second sections are added if warranted by passenger loads.

The opening of light rail service was accompanied by revised connecting bus services. Changes to the bus network were essential to provide access and connectivity to the rail service to fully realize the benefits of an integrated bus/rail service. Bus routes have been restructured so that buses connect with trains at 17 of the 25 light rail stations. Exclusive multimodal transit facilities (transit centers) have been constructed at Coliseum, Hollywood, Gateway, and Gresham Central stations. Bus/rail connections at other stations are made on the street.

Gateway Station is the most critical point of connection between buses and MAX. Timed-transfer operations occur there, with trains and buses pulsing every 15, 30, or 60 min. Inbound and outbound trains pass at Gateway during the timed-transfer "window" in order to make complete bus/rail meets. Timed-transfer operations are also scheduled at Gresham Central, 188th Avenue, 122nd Avenue, and Hollywood stations, particularly during periods of long headway operation. Trains are also scheduled for night and Sunday/holiday downtown meets.

RAIL OPERATIONS PLAN

One of Tri-Met's goals is to operate MAX safely, reliably, and efficiently and to integrate the rail line's operation with bus services for the greatest convenience to the public. The rail operations plan is designed to further this goal by providing information and by documenting procedures and policies necessary to activate and operate the light rail line in the safest, most reliable manner.

Formal development of the rail operations plan began in fall 1985, approximately 1 year prior to start-up. However, the first efforts directed towards conceptualization of the plan date back to 1980, when estimates for staffing plans, operating plans, and operating budgets were developed by a joint venture team of Parsons Brinckerhoff Quade & Douglas, Inc., and Louis T. Klauder & Associates (PB/LTK). Tri-Met began recruiting key rail operations staff then as well.

There was a transitional component to staffing and recruiting for the various phases of the overall light rail project. As the project shifted from planning to preliminary engineering, continuity was maintained by including some of the planners on the newly formed in-house engineering team. Specific technical expertise needs were addressed either by hiring outside talent or through consulting contracts. This strategy built a strongly qualified engineering team, yet maintained the needed links to both the history of the project and Tri-Met in general. Likewise, the same type of transitional staffing efforts followed as the project enlarged in scope to include final design, construction, and operational readiness elements.

In developing the rail start-up organization, Tri-Met's executive management placed top priority on defining the rail operations organizational structure. After various organizational structures from other transit systems with bus and rail modes had been reviewed and analyzed, separate departments for rail transportation and rail maintenance were created in Tri-Met's operations division. With this decision in place, executive management recruited the two key rail operations directors (one promoted internally and one hired from the outside, reflecting a balanced strategy) almost 5 years before actual start-up. Thus, the rail transportation and rail maintenance directors participated in the engineering team's planning and design efforts.

With the engineering project staff working closely with rail operations management, executive management addressed the issue of how to coordinate and prepare the entire agency for start-up. Again, various rail start-up organizational alternatives were reviewed and analyzed; ultimately it was decided to create an interdisciplinary rail operations start-up team. The core of this team was a small group of Tri-Met staff from planning and operations, plus two on-site consultants provided through a rail operations readiness contract with the firm of ATE, Inc. This start-up core team had three key aspects. First, the team members were fully reassigned to lead the start-up effort. Second, the two rail directors were not on the core team in recognition of the greater need for them to continue working closely with engineering.

Third, the core-team leader was formally recognized and authorized by executive management by creation of a director of rail start-up position.

The rail start-up core team was charged with developing the rail operations plan and a complementary start-up activities schedule. Addressing the need for *thorough coordination throughout the entire agency*, the core team identified 14 different functional areas related to start-up as shown below:

- Rail transportation;
- Rail maintenance;
- Safety;
- Security;
- Fare collection and structure;
- Hiring and staffing;
- Information systems;
- Financial forecasting;
- Rail budget development and cost control;
- Marketing and customer services;
- Press, political affairs, and community relations;
- Bus operations;
- Service design; and
- Handicapped access.

Each functional task area was assigned an appropriate task manager, who was responsible for developing the plan and schedule for that particular function. Through a series of weekly coordinating sessions, with all task managers present, the operations plan was refined and revised as necessary, until all task plans were consistent and coordinated.

The formation of a start-up team and the requirement to develop a detailed start-up plan not only provided a working structure for the large coordinating task, but also aided the transition of rail transportation and rail maintenance functions into operating departments. Staffing plans, operating plans, and operating budgets were all reviewed and updated from the preliminary estimates prepared in 1980 by PB/LTK (2). Tables 1 and 2 include PB/LTK's 1980 estimates of light rail operating statistics and costs.

Many of the important elements of the staffing and operating plans, such as the operator's rule book, maintenance rule book, standard operating procedures, training programs, and supplemental agreement to the existing labor contract, were being developed before the start-up plan was commissioned. However, with the additional resources dedicated in the form of a start-up team, it was possible to expedite individual efforts and place them into a cohesive framework. It was particularly advantageous to assign the experienced rail start-up professionals (the two ATE consultants) specifically to the rail transportation and rail maintenance directors.

TABLE 1 MAX LIGHT RAIL OPERATING STATISTICS COMPARISON

Operating Statistic	PB/LTK Estimate 1980	Tri-Met Estimate 1986	Tri-Met Actual 1987[a]
Annual boarding rides (millions)	9.2	4.1–4.9	7.2
Park-and-ride spaces	2,043	1,602	1,602
LRVs	26	26	26
Speed (mph)	19.6	17.1	15.5
Annual car miles (millions)	1.415	1.038	1.286
Annual car hours	72,000	60,000	89,000
Car hours/train hours	1.48	1.32	1.72
Staff			
LRV operators	32	34	34
Other transp/fare inspection	26	20.5	20.5
Vehicle maint/stores	28	26	26
ROW maintenance	23	25.5	25.5
Total staff	109	106	106

[a]September 1986 to August 1987.

TABLE 2 MAX LIGHT RAIL COST ESTIMATE COMPARISON

Annual Operating Costs	PB/LTK Estimate 1980	Tri-Met Estimate 1986	Tri-Met Actual 1987[a]
Rail transportation	3.002	2.085	1.894
Rail maintenance	3.626	3.144	2.558
Electrical power	1.331	0.840	0.567
Insurance & claims	0.167	0.168	0.092
General & administrative	0[b]	0.987	0.889
Estimated annual cost	8.126	7.224	—
Actual annual cost	—	—	6.000
Cost/car mile ($)	5.74	6.96	4.67

NOTE: All operating costs are in millions of 1987 dollars.
[a]September 1986 to August 1987.
[b]G&A costs included in rail transportation and maintenance figures.

In recognition of the importance of the peer review process, the rail operations plan called for continuing and intensifying the process initiated with the first peer review held in September 1984. Thus, additional peer reviews were held in February and August 1986 (1 month before start-up). Also, at Tri-Met's request, a system safety review was conducted by the American Public Transit Association's Rail Safety Review Board. All of the peer reviews provided excellent recommendations to improve Tri-Met's LRT system.

As it was being developed, the rail operations plan represented a vision of what the end product should be, namely a safe, reliable, efficient, and integrated light rail line. The companion volume to the rail operations plan, the start-up activities schedule, represented the process for achieving the goals enumerated in the plan.

START-UP ACTIVITIES SCHEDULE

The purpose of the start-up activities schedule was to summarize the sequence and timing of all activities required to establish revenue service on the target date, September 5, 1986. The schedule was actually a series of separate schedules that described the event sequence and deadline dates for each of the 14 task areas identified in the rail operations plan.

The first set of activity schedules was issued in December 1985, concurrent with the production of the second monthly progress report on the start-up effort. Each subsequent month, a new set of schedules, updated and reflecting progress made, was issued together with the monthly progress report up until the September 1986 start-up date.

In selecting a format and methodology for the activities schedule, various computerized and manual systems were analyzed. Ultimately, the start-up core team chose to use a simple manual tracking chart of a simple matrix design, with rows identifying tasks and subtasks, and columns denoting time in monthly gradations. This approach was selected because it maintained continuity and familiarity by replicating the engineering activities scheduling system, and maximized the simplicity and comprehensibility of the project scheduling system, particularly for nontechnical team members.

The core team was also concerned that team members might think that a detailed automated project scheduling system would obviate the need for oral project communication. Thus, the strategy was to foster open, face-to-face communication, in part, through the weekly coordinating meetings, and to position the easy-to-use activities schedules as supporting documents, useful for task monitoring and accountability purposes.

In many functional task areas the individual activities schedules were fairly straightforward and almost perfunctory in nature. However, there was one critically important start-up task that benefited from the development of activities schedules: the rail operations recruitment and training program. By graphically identifying subtask time requirements for recruiting, testing, training, and appointing different classifications of operating personnel, it was possible to develop a comprehensive, incremental schedule for staffing rail operations. The incremental schedule was then adjusted periodically to match the engineering staff's updated construction and equipment testing schedules, so that operating staff appointments coincided with the availability of equipment and facilities for training purposes.

Based upon the results of the supplemental working and wage agreement relating to light rail operation negotiated with union representatives, preference was given to qualified Tri-Met employees when filling positions for LRT operations and maintenance. The agreement also stipulated that all normal work would be performed by Tri-Met employees. Outside contractors could only be used for emergency repairs, unanticipated work overloads, and specialized heavy-duty maintenance for which Tri-Met does not have the necessary equipment.

For the rail transportation department, this meant that the rail controller and supervisor positions would be appointed according to seniority from the ranks of qualified bus supervisors. Similarly, light rail operators would be appointed according to seniority from the ranks of qualified bus operators.

The process for appointing rail controller/supervisors and operators was very thorough. It included personnel file reviews (with acceptable performance levels identified), written examinations of the rail operator's rulebook, medical examinations, and, after acceptance into the training program, daily written examinations and quizzes. Even with this relatively straightforward approach to staffing and training, various complexities surfaced, including coordinating replacement supervisors and operators for bus operations, separating total staffing complements into subgroups for effectively sized training classes, and rescheduling tasks based on replacement candidates' availability.

The rail maintenance department was organized into two sections: vehicle maintenance and right-of-way maintenance. For the vehicle maintenance section, foreman, LRV mechanic, and fare/lift equipment maintainer positions were appointed according to seniority from the ranks of qualified bus maintenance employees. The same agreement was in place for staffing the right-of-way section, which led to the appointment of rail right-of-way maintainers and cleaners from the bus maintenance building and grounds section. However, for the various skilled right-of-way labor positions (power maintainers, signal maintainers, etc.), in-house, qualified candidates were few. Thus external recruitment was required. Also, in some cases, qualified applicants were transferred from the Banfield Light Rail Project engineering department to the rail right-of-way maintenance section. The development and use of a simple, flexible activity schedule for coordinating and tracking the complexities of staffing the rail transportation and rail maintenance departments were quite helpful.

COST ESTIMATES AND RESULTS

Tri-Met's light rail operating cost estimates originated with the work performed in 1980 by the joint venture of Parsons Brinckerhoff Quade & Douglas, Inc., and Louis T. Klauder & Associates (PB/LTK). These cost estimates were documented in their Phase II report (2). The report was one of 11 technical reports that dealt with specific elements of the project.

Tables 1 and 2 compare 1980 estimates of operating statistics and costs with Tri-Met's 1986 estimates and actual results for 1987. PB/LTK's cost figures were originally calculated using 1978 dollars, then factored up to 1980 dollars using an 8 percent annual rate. For comparative purposes, these costs have been factored back to 1978 dollars and then multiplied by the actual annual change in the U.S. Consumer Price Index to determine equivalent costs in 1987 dollars.

PB/LTK's cost estimates were developed from an operating scenario that estimated first-year ridership of 9.2 million boarding rides. (This represented a level of service about one-third higher than the revised May 1986 Tri-Met estimate.) Based upon this service level, PB/LTK determined staffing and materials requirements.

Staffing estimates were based upon the organizational structures of other transit properties and Tri-Met's labor practices and productivity rates. Staffing assumptions were considered adequate for regular maintenance activities, with some contracting for specialized, heavy maintenance activities (track rebuilding, rail grinding, etc.). Power costs were based upon private utility company rate structures. PB/LTK estimated annual light rail operating and maintenance costs of approximately $8.2 million and assumed no increase in bus operations or administrative costs. A rail operations staff of 109 was estimated to be required to provide 1.415 million annual car miles of service.

Beginning in autumn 1985, Tri-Met tried to refine PB/LTK's original estimates and assumptions. Numerous iterations resulted in May 1986 estimates that included a staff of 106 providing 1.038 million car miles of service. The first-year ridership estimate was substantially reduced to a range of 4.1 million to 4.9 million boarding rides. The 1986 Tri-Met annual operating and maintenance cost estimate was 11 percent lower than the 1980 PB/LTK estimate. The estimated operations and maintenance cost per car mile is $6.96, compared with PB/LTK's estimate of $5.74, because Tri-Met reduced PB/LTK estimated operating speed by 2.5 mph (to 17.1 mph). The fairly sharp changes between the PB/LTK estimates and the Tri-Met figures are due primarily to the 6 years that elapsed between the two sets of assumptions underlying the estimates. Prior to 1986, Tri-Met developed several updates to PB/LTK's 1980 cost estimates; however, until the start-up coordination team was in place, in-house efforts to update operating assumptions and cost estimates were difficult.

First-year actual results (September 1986 to August 1987) reflect the higher-than-anticipated ridership level, as well as the additional car hours of service required to support this ridership level. The car-hour/train-hour ratio is higher than previous estimates, reflecting the need to operate more two-car consists as ridership levels warrant. The operating speed is considerably less than expected, due to slower-than-planned operating speeds along Holladay Street and the downtown Portland alignment.

First-year costs are $1.2 million below the Tri-Met 1986 estimate, because actual power and maintenance costs were significantly under budget. Power costs are expected to remain relatively stable at this favorable rate over the next few years. However, rail maintenance costs are expected to increase gradually in the next 2 years as the system ages and as warranty agreements expire, necessitating additional in-house labor resources. Beyond the 2-year mark, rail maintenance costs should remain stable.

SUMMARY

A successful light rail start-up project requires a strong commitment from executive management to create a start-up core team by contracting with experienced start-up consultants and fully reassigning key staff, and to support the leader of the team by conferring both the authority and resources required by the project.

A comprehensive rail operations plan and a complementary start-up activities schedule are essential project control documents. The primary purpose of producing and regularly revising these documents is to provide start-up team members with reference materials during weekly and daily communication and coordination meetings.

The large scope of a rail start-up project requires that considerable energies be focused towards resolving a myriad of detailed issues. To keep the overall project priorities in place and continuously synchronized, it may be useful to develop a start-up summary checklist. This checklist, or set of guidelines, can be drawn from lessons learned and experience gained by other properties' rail start-ups. An excellent method to assist in developing a property-specific set of guidelines is to conduct peer reviews at regular intervals.

REFERENCES

1. T. Matoff and K. Zatarain. *On and Around MAX: A Field Guide to Portland's Light Rail System*, 2nd ed. Tri-Met Transit Development Department, Portland, Oreg., November 1987.
2. P. H. Gilbert and L. T. Klauder. *Banfield Light Rail Project Pre-Design Report*. Technical Memorandum No. 10, Capital Cost Estimates/Operations & Maintenance Costs. Parsons Brinckerhoff Quade & Douglas, Inc./Louis T. Klauder & Associates, Portland, Oreg., July 1980.

RT Metro
Trials and Tribulations of a Rail Start-Up

CAMERON BEACH

Putting Sacramento's RT Metro on the track took years of coordination and cooperation among government bodies, internal departments, contractors, and vendors. A core management staff was brought aboard early to plan the system's eventual operation. Governmental bodies other than the system operator itself, Sacramento Regional Transit (RT), eventually bowed out, giving RT full responsibility. City redevelopment funds were tapped to make up a shortfall in the original budget. Well in advance of the opening of the system's first leg in March 1987, RT managers negotiated with labor unions, assembled an Operations Coordination Committee as a liaison with law enforcement and fire department officials, and established a training program. As sections of track were turned over for testing, extensive walk-throughs were done, followed by further testing using a light rail vehicle (LRV). The LRVs themselves were tested extensively and a video camera mounted on top of one of them was used to check catenary construction and wire stagger. About 3 months before the system opened, simulated revenue service was begun to make sure the system would operate as expected.

AFTER 10 YEARS OF PLANNING and 5 years of construction, the Sacramento Regional Transit (RT) District opened the first 9.5-mi leg of RT Metro on March 12, 1987. The second leg, completing the 18.3-mi starter line, opened September 5, 1987.

Regional Transit was not initially responsible for construction of the line. A joint-powers agreement was signed with the City of Sacramento, the County of Sacramento, Caltrans, and Regional Transit to form the Sacramento

---

Regional Transit, P. O. Box 2110, Sacramento, Calif. 95812.

Transit Development Agency (STDA) in 1981. STDA's mission was to design, engineer, and construct a light rail system for Sacramento that would be turned over to Regional Transit for operation upon completion.

Early on, RT's senior management recognized the pitfalls of having a system designed and constructed without extensive input from the operator. With the line scheduled to open in spring 1985, RT General Manager David Boggs appointed a light rail manager in July 1983. This manager would be responsible for putting together a start-up plan, which would include the hiring and training of all employees. The manager was also responsible for coordinating design and construction activities with STDA.

By mid-1984, it became apparent that the spring 1985 opening date was not realistic. During this time, it also became apparent that the $131 million budget was not sufficient to construct the system as designed. Because RT had the financial responsibility for completing the project, it was decided that the construction responsibility should be RT's as well.

On August 15, 1985, the STDA was dissolved and responsibility for the light rail project fell solely on RT. At that time, a more realistic budget of $159 million was adopted utilizing city redevelopment funds to make up the difference.

During these times, RT's operations group put together a staffing plan that called for 68 employees to operate and maintain the light rail system. While staff felt that the number was low, budgetary considerations did not allow for higher staffing levels. In early 1985 the transportation superintendent and the maintenance superintendent came on board. The transportation superintendent authored the first draft of an operating rulebook. The maintenance superintendent was kept busy coordinating design reviews with the vehicle manufacturer, traction power installer, signal installer, and trackwork contractors.

The staffing plan was amended numerous times, primarily as the result of input from peer reviews conducted in 1985 and 1986. Primary increases in staffing occurred in wayside maintenance, fare inspection, and, to a lesser extent, in vehicle maintenance. The maintenance staff today is able to keep up with RT's requirements, but as the system gets older, additional personnel will be necessary.

During 1985 and 1986, negotiations were conducted with the Amalgamated Transit Union (ATU) and the International Brotherhood of Electrical Workers (IBEW) regarding wages and working conditions for the light rail operations employees who would be members of the respective bargaining units. The ATU represented the bus operators and office clerical staff, while the IBEW represented the maintenance employees. It is interesting to note that the IBEW representation came about because RT's predecessors operated the streetcar system in Sacramento until its abandonment in 1947.

Issues discussed with the unions included methods for selecting train operators and maintenance personnel, job descriptions, wages, and representation of new classifications. Visits were made to other rail operating properties represented by the ATU and IBEW to compare job duties and provide insight for union representatives who had previously only dealt with bus-related issues. Negotiations with the unions were concluded in late 1985 with the signing of side agreements to the existing contracts.

All work on the system is done by RT's own employees with the exceptions of station cleaning, landscape maintenance, security, and weed abatement, which are contracted out to local firms. When they become necessary, tasks such as traction motor rebuilding will also be contracted out.

Another area that required a great deal of attention was coordination with the police and fire departments. Because streetcars had been gone from Sacramento for over 40 years, the whole idea of overhead wires in the middle of a street was foreign to fire-fighting personnel. In addition, law enforcement officers needed additional training on how to deal with trains operating in traffic on the street.

To address this problem an Operations Coordination Committee was created. It consisted of police officers, deputy sheriffs, highway patrol officers, fire chiefs, and their respective training personnel as well as RT rail operations staff. This group met every other month for almost 2 years prior to the system's opening. Numerous questions and issues were raised during these meetings. There is no doubt that the current good working relationship with these groups is due to these efforts.

Operations staff moved into the Metro Division Operations and Maintenance Facility during early November 1986. The first light rail vehicle (LRV) was delivered 2 weeks later. To have properly trained personnel to operate the cars, the transportation superintendent, the two senior transportation supervisors, and the two most senior train operators were sent to Calgary, Alberta, for extensive training in LRV operation and train control. To this day, these individuals talk about their "vacation" in Calgary. All but one of them were native Californians who had a difficult time adjusting to the −15°F to −35°F temperatures that they encountered in Alberta in November.

It is important to note at this point that over 85 percent of RT Metro's employees were promoted from within the ranks. RT made a commitment early on that an expansion into light rail would mean new opportunities for existing staff. Only those positions that required specific technical expertise were filled from outside the agency. Prior to sending staff to Calgary, bus operators were asked to sign a list indicating their interest in light rail training. At that time, the level of interest that would be expressed was unknown. But after 1 week 175 of RT's 320 bus operators indicated they wanted to learn to run trains. The two operators selected to go to Calgary had almost 60 years of cumulative experience driving buses for RT and its predecessors.

The process in which the construction department turned over areas of track for testing and operation took a great deal of effort and patience on everyone's part. Sacramento was fortunate to have a high level of cooperation and camaraderie between construction and operations personnel throughout the project. Without this, it is questionable whether the system would ever have worked. Extensive walk-throughs were held on all phases of construction by operations personnel. Prior to any testing, every foot of track and overhead was inspected by operations staff. Following successful completion of this last walk-through, an LRV would be moved at no greater than walking speed through the affected territory. Speeds would be increased in 5- or 10-mph increments until track speed was reached. This process was slow and tedious, but in one case it prevented an LRV from hitting a curb that was too high and pointed out such problems as trees growing into the overhead.

Prior to train operations on the test track, procedures were developed for test track limits and "red tagging" of traction power so that both contractors and testing crews could work simultaneously. The buffer zones between construction and operations were established with track warrant and red tag procedures being rigidly enforced by operations personnel.

An extensive testing and burn-in program was developed for testing and accepting LRVs. Our first two operators made so many trips over the original 1.5-mi test track there were days they felt like they were operating a horizontal elevator.

As a part of their training, operators had been instructed that a dark (unlit) signal must be treated as a "red" signal. As signal equipment was installed by the contractor, burlap bags were used to cover the signal heads so that operators did not have to disobey operating rules.

As longer sections of the system were completed, a formal program was instituted that provided for extensive testing of the system and its components prior to unlimited use by operations. Once the walk-through and slow running tests were completed, a video camera was mounted on top of an LRV to check catenary construction and wire stagger. Following this, extensive system tests were conducted of each of the components, i.e., signals, switch machines, substations, traffic signals, and dynamic clearances. For example, each signal was checked for visibility from an operating cab. Each possible routing was checked to verify proper signal aspects and prevention of conflicting moves. Each substation was load tested by having two fully loaded four-car trains accelerate away from each other on a single feed.

Once this was accomplished, an extensive series of integrated tests was conducted to determine that all of the subsystems worked together properly. This included radio coverage tests, platform measurements with an LRV to verify clearances, and timing of traffic signal preemption devices to optimize train movements.

About 3 months prior to the opening of the system, an extensive program of operational testing was begun. Called simulated revenue service, this was the final test of whether the light rail system would operate as the planners and engineers intended. The system was designed for eight trains to operate on a 15-min headway. A computer simulation compiled by Foster Engineering of San Francisco showed that such an operation was possible, but that meets would be close on some stretches of single track. A plan to include additional sections of double track was deferred by the board until actual operation confirmed the need for the expenditure of additional (and very scarce) capital funds.

Included in simulated revenue service were the final aspects of operator training, for example, simulated and actual passenger boardings, elderly and handicapped access, schedule adherence, train meets, cuts and adds to train consists, and verification of running times previously plotted by computers. In addition, several incidents were staged to test the acuity of both the transportation and maintenance personnel as well as various public safety agencies. These incidents included derailments, collisions, signal failures, and other disruptive activities.

During the final 2 weeks prior to opening, a multialarm fire was to be simulated on the K Street Mall during the afternoon rush hour. The purpose of this test was to determine RT's ability to respond to such an incident and maintain an appropriate level of service. But 2 days before the test was to occur, there was an actual multialarm fire on the K Street Mall. The Sacramento Fire Department was able to utilize specifically created procedures for shut down of traction power and protection of fire fighters. RT personnel were able to test their ability to cope with a major disruption to service without actually affecting the riding public.

The simulated revenue service testing proved to be an unqualified success. Without this herculean effort, it is doubtful that the March 12, 1987, opening of Sacramento's light rail system would have come together so well.

Had we the opportunity to go back and do it again, relatively few things would have been done differently. Operations input into the signal system design would have been greater. Signal design was based on the Association of American Railroads standards. As an example, a green-over-red aspect would be displayed at a diverging route. This wasn't a problem for the system's ex-railroaders, but bus division employees had been taught never to go past a red signal. The green-over-red became an exception to the "never" rule. Therefore, in the interest of uniformity, the aspects and heads were reworked to provide that any red aspect would require permission from Metro Control before proceeding. In addition, we would have insisted on more comprehensive training from the signal and traction power contractors, similar to what was provided by the vehicle manufacturer.

During the construction and start-up process there were many occasions when operations staff, engineers, test crews, and contractors became frustrated with the whole process. But without the great deal of cooperation and attention put forth by these groups, Sacramento's system would not be where it is today.

Preparation for "Show Time"
The Los Angeles Story

Norman J. Jester

The Los Angeles-Long Beach Light Rail Transit Project is scheduled to begin revenue operations in 1990. With the $700-million, 21-mi project already under construction, careful planning and coordination are being conducted to ensure that the system will give a stellar performance on the day of its debut. A test track will be constructed so that vehicle acceptance can proceed uninterrupted while construction of the system continues. Staffing is an area where early planning has already begun to pay off by providing an accurate picture of first-year operational expenses and by establishing an incremental plan to bring on staff just in time to meet the needs of each development phase. Planning for the training of operating and maintenance employees is also under way. And a detailed testing plan has been put together to verify the system equipment at every stage of development. Other considerations include labor agreements, logistics of spare parts and supplies, public relations and marketing, contingency plans, maintenance vehicles and shop facilities, and safety certification.

THE LOS ANGELES-LONG BEACH Light Rail Transit (LRT) Project is the forerunner of a projected multiline light rail network that will, it is hoped, someday encompass the entire Los Angeles basin. The initial line (see Figure 1), running some 21 mi from downtown Los Angeles to Long Beach and making 22 stops, is scheduled to open in July 1990.

Los Angeles County Transportation Commission, 403 W. Eighth Street, Suite 500, Los Angeles, Calif. 90014.

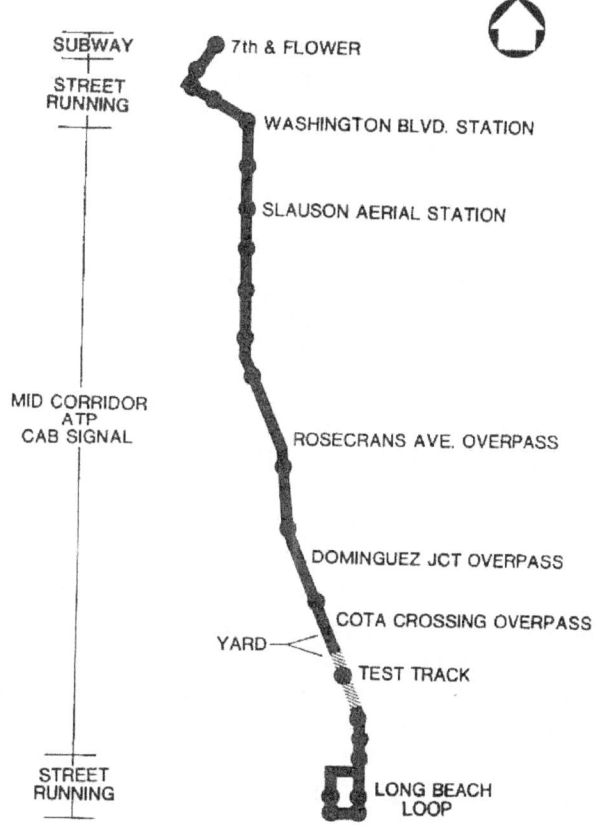

FIGURE 1   Los Angeles-Long Beach LRT schematic.

The Long Beach line will use conventional, state-of-the-art light rail technology: driver-operated articulated rail cars moving largely over surface tracks, interfacing with automobile traffic at numerous grade crossings, and including approximately 6 mi of street traffic operation. There will also be about 1 mi of subway operation in downtown Los Angeles, plus several short stretches of elevated structure serving as flyovers across congested areas.

Stations will be high-platform. Train-operator controlled track switches and a cab-signal system will govern train movements on areas of private right-of-way. A supervisory control and data acquisition system (SCADA) will allow central control dispatchers to monitor and control the traction power supply system, monitor train movement and ticket vending machines, and operate some interlockings during emergencies. Street-running and yard movements will be governed by an operations rule book.

The fare collection system will be the self-service system that has been successfully demonstrated in at least seven North American cities to date.

Two-car trains will predominate in the operation, running on 6- to 10-min headways in peak periods, 15- to 20-min headways off-peak. Initially, the light rail vehicle (LRV) fleet will number 54 and will be constructed by the Nippon-Sharyo works in Japan.

The cars will be delivered in "halves" and assembled in the newly constructed maintenance shop in Long Beach.

Funding for this $700-million project is entirely local, much of it coming from a 0.5 percent sales tax locally referred to as "Prop A" funds after the proposal was enacted in early 1981. Design of the system began in earnest in early 1985 and ground-breaking occurred October 31, 1985, at the Long Beach shop site. Much of the route alignment is shared with the Southern Pacific Railroad, which must maintain its freight service during the 4½-year construction period—not an easy task.

Although 4½ years may sound like a lot of time to spend building track on an existing right-of-way, when one takes a closer look at all of the individual pieces of the giant puzzle that must mesh together, time is of the essence if the trains are to roll on the scheduled opening day.

Few projects of this magnitude (aside from space exploration) involve more disciplines plus the public than a rail transit project. The Los Angeles project ranks among the biggest and is without question the largest single light rail project in this country. The list of players, all of whom have a role in opening the system, is awesome. Architects, planners, schedulers, engineers, draftsmen, politicians from many diverse constituencies, administrators, clerical workers, technicians, public relations persons, equal employment opportunity (EEO) officers, contractors, procurement specialists, insurance specialists, legal counsel, journeymen, utility companies, real estate agents, developers, parking authorities, traffic engineers, the state Public Utilities Commission, safety experts, police and fire jurisdictions, newspaper reporters, tax authorities, vending machine salesmen, the local bus operating agencies, test coordinators, operator training specialists, and on and on—all play active roles. Getting all of these people and organizations to work together towards the same goal is tantamount to getting a large army of fleas to march in step.

Yet by June 1990 the taxpayers of Los Angeles will have been paying sales taxes for 9 years and for 9 years they will have been reading about the promised beginning of the end of their gridlock nightmares. For over 4 years they will have been negotiating construction disruption in downtown LA and elsewhere. For as long as they can remember, they have been fed promises of this great new set of trains that will whisk them to their jobs conveniently. If the day arrives and the trains are not there, the taxpayers will vent their wrath in the press and at the polls. Our duty is not to let the public down when the big day arrives. That big day is what we are referring to as "Show Time."

To make all this happen, the planning must begin early in the design stage of the project. Actually it begins with the schedulers who determine which segment of the line has a possibility of being operable at the earliest date. Presently the Los Angeles County Transportation Commission (LACTC) is planning to start the system in two stages. The first stage, opening in July 1990, includes the entire system except the subway section, which is scheduled to open in December 1990.

Our story of getting ready for "Show Time" begins then in early 1985 when the final Environmental Impact Statement was approved, the route alignment was finally chosen, funding was in place, and a design engineering consulting team was brought on board to do the planning and design work.

TEST TRACK

From the very early days of the project, the need for a test track was recognized so that vehicle acceptance could proceed uninterrupted while construction of the system continued. The area selected for the test track is between the yard departure track and the Willow Station pocket track, a distance of approximately 10,300 ft. There are two grade crossings on the test track, but there is sufficient distance between the crossings for high-speed (55-mph) performance testing of the LRVs.

To ensure that at least the minimum requirements for a test track will be completed in time to support vehicle acceptance, schedule milestones were included in all the system contracts and some construction contracts. A special effort will be made by all our resident engineers to make sure that the test track work proceeds expeditiously so that vehicle acceptance will not be delayed. This includes the portions of the yard and shops required to support the vehicle deliveries.

STAFFING

Another area where LACTC and its consultant staff initiated early planning was staffing. Besides giving us a very accurate picture of what operating expenses will be for the first revenue service year, this early planning will also allow smooth transitions from the testing phase through the prerevenue and revenue service phases of the project. The staffing plan for the system is incremental throughout the test and start-up phase, with personnel being brought on board only as needed to support testing, to ensure sufficient training time, and to provide preventive maintenance on equipment already accepted.

TRAINING OF PERSONNEL

A well-rounded training program is of key importance to the timely opening of the system and to ensure the safety of our passengers during the initial stages of revenue service. For the purposes of these discussions, our employees can be divided into two categories, operating employees and maintenance employees. We will not discuss the training of supervisory or specialized personnel, such as dispatchers, because these people will be hired as experienced personnel, and will only be required to familiarize themselves with the new environment and equipment.

Training of operating employees (see Figure 2) will consist primarily of two separate classroom sessions, with a lot of hands-on training in operating the vehicles before and after each session. The testing period will be an essentially hands-on training opportunity for the operators.

The operators will be given two written tests during their training period. The first will cover safety rules and those portions of the rules that pertain to test operations only. The second will involve the entire operating rules and procedures manual. This is done to avoid confusion, because operators must perform test operations with some rules and procedures that will not apply until prerevenue tests begin. The prerevenue test period is considered the final for operating personnel. During this period the operators' adherence rules and procedures, their reactions, and their responses to emergency and abnormal conditions will be evaluated and graded. Those requiring additional training, or those unsuitable for continued service, will be identified.

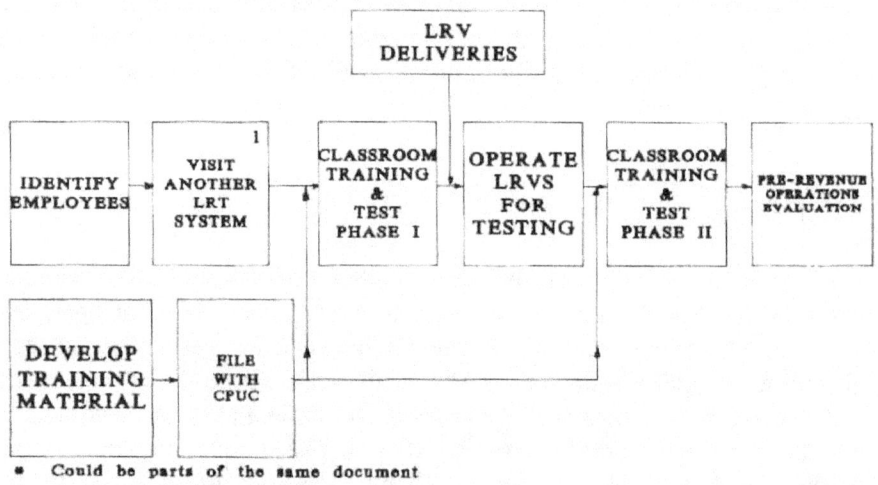

* Could be parts of the same document
1 Applies to initial group only

FIGURE 2 Training phases for operating employees.

The maintenance personnel (see Figure 3) hired for skilled positions will be required to possess at least basic skills. If the decision is reached to make the hiring of existing personnel a priority, some basic skills training may be required, particularly in electronics. This basic skills training will have to be coordinated with various technical schools in the area. All of the system's equipment specifications include requirements for vendors to provide classroom and hands-on maintenance training for our employees.

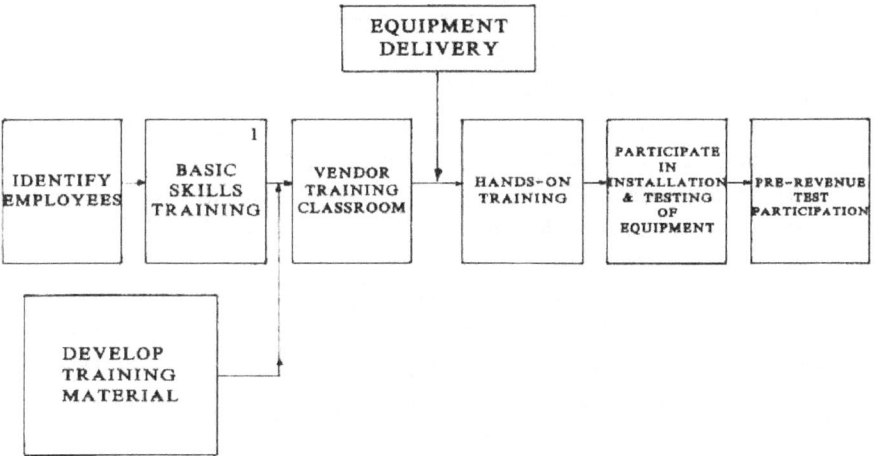

1 May be required if present RTD maintenance personnel are used for light rail system

FIGURE 3 Training phases for maintenance employees.

Maintenance personnel will also be used to monitor tests and will participate under the direction of the contractor in any retrofit or repair work on the equipment. This will provide our maintenance personnel the opportunity to gain further experience with the equipment.

TESTING OF THE SYSTEM

Test planning also started very early in the program and a very complete test program has been devised to accomplish four goals:

1. Demonstrate the safety and service characteristics of the system;
2. Validate and demonstrate system performance;
3. Verify contract compliance; and
4. Train personnel and integrate personnel, equipment, and procedures.

Test sections of the various contracts have been very carefully written to require the contractor to verify the equipment at every stage of development, starting with materials and manufacturing tests, the installation verification tests, and finally the integrated tests to verify that all systems work together as a fine-tuned light rail system.

Figure 4 is a flow chart of the field testing that will be taking place in a typical section of the midcorridor. Only tests that will be performed in Los Angeles are shown, but factory or qualification tests to be done at the factory are excluded. Numbers in parentheses in the following discussion correspond to the numbered tests in Figure 4. Starting from the left, the vehicle acceptance test (1) is performed at the test track on every vehicle received. After each LRV successfully completes this test, it will be used on integrated tests on other sections of the system. The vehicle test involves the verification of performance requirements in the specifications (such things as acceleration and deceleration rates, brake cylinder pressures, motor currents, etc.). The automatic train protection (ATP) and train-to-wayside control (TWC) vehicle equipment will also be tested to verify proper operation with the wayside installation.

After accepted vehicles are available, clearance tests (2) of the particular section under test will be performed. This test involves fitting a vehicle with a simulated dynamic profile to verify that proper clearances exist at all the physical facilities on the right-of-way, such as poles, switch machines, platforms, etc.

Overhead catenary system (OCS) mechanical and electrical tests (3) are the next step in the testing process. After mechanical checks have been done to each section of the OCS to verify the height and stagger of the contact wire, and after electrical installation verification tests (such as circuit continuity and loop resistance), hi pot insulation tests, and grounding resistance tests are complete, dead wire run tests (4) will be performed on the particular section. This test consists of using an LRV pulled by a rail-mounted vehicle to verify the pantograph sway and pantograph clearance envelope. This test is performed at several speeds in 5 mph speed increments.

Traction power supply system installation tests (5) are conducted at each traction power substation. These include insulation resistance tests, circuit continuity tests, and grounding system tests. The section will then be ready for the energization test (6) to verify proper voltage in the OCS.

Live wire tests (7) will then be performed to evaluate the collection performance between the LRV and the OCS. These tests will be done at various speeds (starting at 5 mph) in increments of 5 mph. Videotape recordings will be made of the interface between the pantograph and the OCS to verify behavior on various contact wire profiles and to show loss of contact, smooth transitions at overlaps, cross contacts, turn outs, and section insulators.

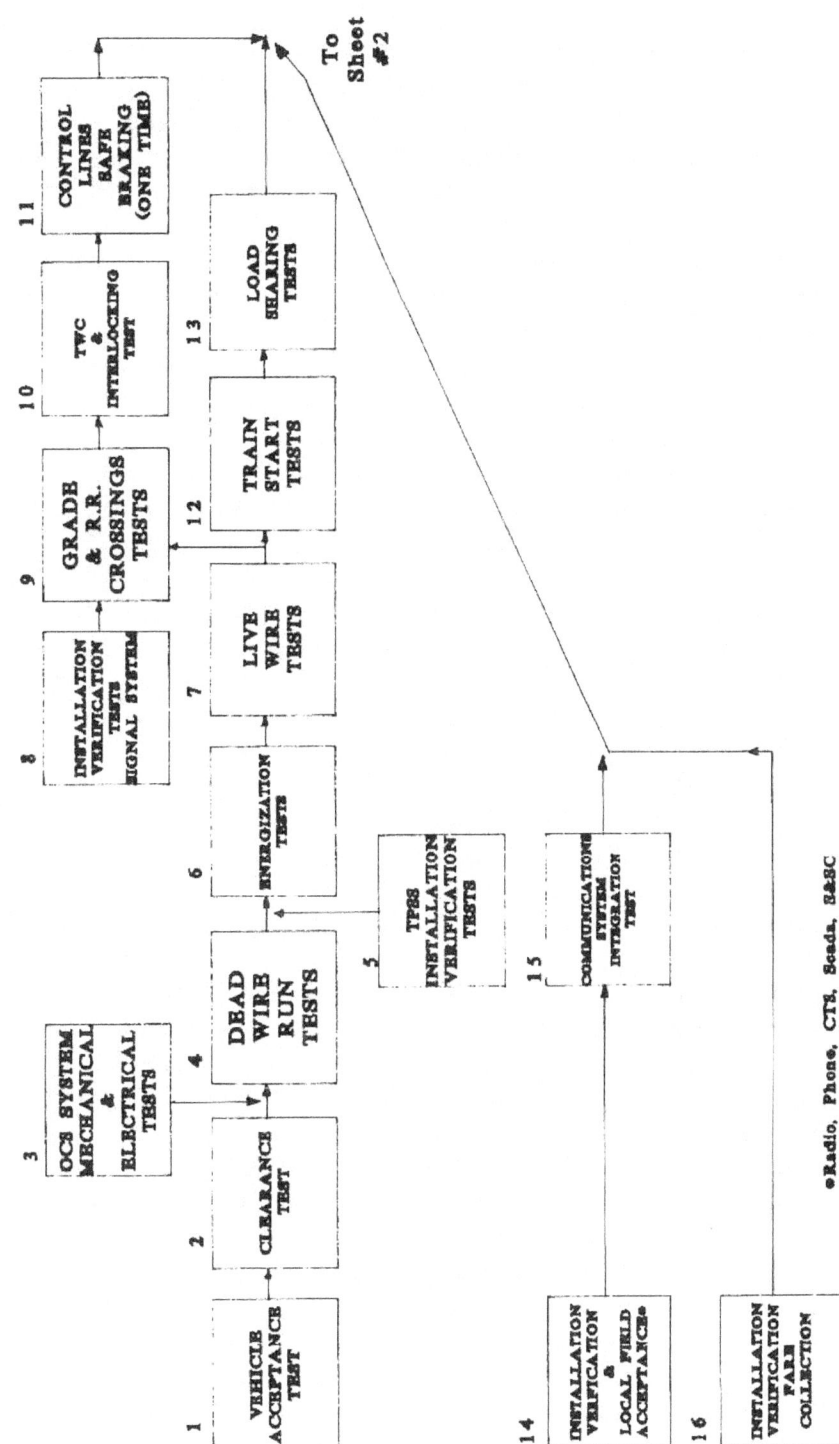

FIGURE 4 Testing phases for LRT components.

The signal system will be tested (8) after the proper installation verification tests of the signaling equipment are complete. These include tests of cable continuity and resistance, power bond resistance, insulated joints, switch machine installation, energy distribution, AFO and ac track circuit adjustment, signal adjustment, line circuits, traffic circuits, TWC, and interlocking verification.

Every grade crossing and railroad crossing will be thoroughly tested (9), first using shunts and then using actual LRVs. Grade crossing protection operation and timing will be verified for all speeds. Railroad crossings will be tested from both the LRT side and the Southern Pacific Railroad side. The TWC operation will be verified at all locations using actual LRVs and verifying every request at each location.

TWC and interlocking tests (10) follow. Each interlocking will be thoroughly tested to verify that the vital circuits, switch locking, detector locking, and signal locking operate as designed. All routes will be checked and route security proven using actual LRVs.

Next come the automatic train protection system tests and the safe braking test (11). Control lines will be tested with actual LRVs to verify that the correct speed command is received in each track circuit for every condition of track occupancy in the block preceding the track circuit under test. The test will be performed with one LRV incrementing from track circuit to track circuit. For one selected track circuit a safe braking distance test will be performed to verify that block design is sufficient to ensure safe train separation. An LRV with derated braking characteristics will be used for this test. These tests, of course, will not be performed in the downtown sections—Los Angeles and Long Beach—but traffic signal sighting tests will be performed there.

At each traction power substation one two-car train-start test (12) will be performed to verify the proper operation and coordination of the dc feeder breakers during the start. Also at each substation the proper load sharing of the two ac-to-dc conversion assemblies will be verified (13) during a three-car train start.

Local field acceptance tests (14) will be conducted on all the communications systems, namely the cable transmission system, SCADA, radio, telephone, public address, closed-circuit television, fire detection and suppression monitoring, and intrusion detection systems. When these are completed, an integrated test of the communications systems (15) will be conducted.

Installation verification tests will be conducted at each ticket vending machine (16). These will verify their proper operation and data reporting to the central data computer as well as the maintenance record-keeping computer.

After every section of the system has been tested as indicated in the preceding paragraphs, complete system tests will be run (see Figure 5).

The capability of the signal system to perform as designed will be verified by performing a system operations test (17). Three-minute headways and a three-hour-long rush hour simulation test (18) will be performed. Some failure modes for the signal system will be verified during this test.

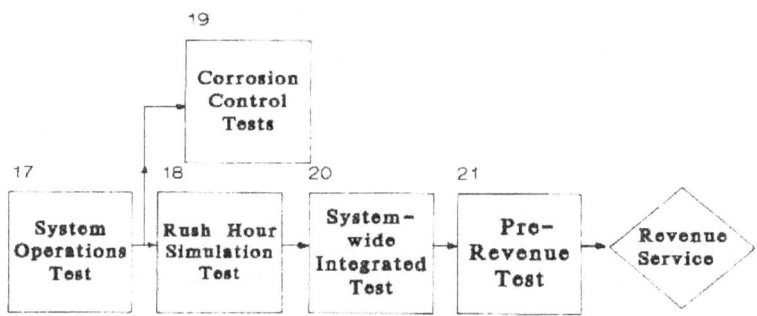

FIGURE 5 Testing phases for systemwide integrated tests.

Corrosion control tests (19) will be performed in conjunction with the simulated rush hour test to verify the effectiveness of the corrosion control measures installed within the transit system.

After all the above testing has been completed, a systemwide integrated test (20) will be performed by the LACTC and its consultant staff. This test will be attended by representatives of all the system contractors. Its purpose is to verify the complete integration of the SCADA, LRVs, signaling, traction electrification system, fare collection, and the various communications systems. The function of every annunciation and control circuit will be verified and all the operating features of the LRT system will be demonstrated.

After the systemwide integrated tests have been completed, the prerevenue operations phase will start. All testing up to this point is intended to verify equipment operation. The prerevenue operations tests (21) are intended instead to verify the knowledge and reactions of people.

There are three kinds of prerevenue operations tests, namely, those for normal revenue operations, abnormal operations, and emergency scenarios. All of the system equipment elements and operating employees will participate in these tests. The performance of employees will be observed and graded. Operating employees include the train operators, their supervisors, dispatchers, the maintainers, and their supervisors. The appropriate emergency response units for all of the areas over which the system operates will be called upon to participate in the emergency scenario portions. Employees who do not perform adequately will be scheduled for retraining or termination. Then, and only then, if all safety certification requirements have been satisfied, the system will be ready for revenue service.

OTHER CONSIDERATIONS

Several other considerations must also be dealt with: labor agreements, logistics, public relations and marketing, contingency plans, maintenance vehicles and shop facilities, and safety certification.

Labor Agreements

In the very near future some important policy decisions have to be made regarding the source of employees for the light rail system. If it is decided that the employees must come from within the ranks, negotiations must begin in the not too distant future. An agreement must be developed under which the operating employees can make the transition from the bus system to the rail system. Such issues as separate rosters for rail employees must be addressed by union and management alike. The trend towards privatization will also be considered. The cost and benefits of having any of the work performed by available private contractors must be known prior to any negotiations.

Logistics of Spare Parts and Supplies

All the systems contracts contain provisions for the equipment contractor to provide spare parts and special test equipment. They also must provide a suggested list of consumable and special tools. Decisions on the types and quantity of consumable supplies and tools need to be made in the near future. Work on the layout of the storeroom, presently under construction, also needs to be initiated.

Public Relations and Marketing

Some brochures have been prepared in English and Spanish to inform potential patrons and others about the LRT. An additional concentrated effort must be made through the media prior to revenue operations to ensure proper exposure to all our potential riders. Decisions must be made about whether demonstration rides will be given during the testing period (i.e., at the test track) or if a period of free rides will be instituted prior to revenue service.

Contingency Plans

During the entire process, contingency plans are constantly being formulated to cope with construction delays in certain line sections and facilities. Some contingency plans have also been made for opening smaller segments of the system, should it become necessary.

Maintenance Vehicles and Shop Facilities

Various high-rail maintenance vehicles must be purchased, tested, and their operators trained. Because these vehicles will not shunt the signal system and be detected, appropriate rules must be developed and incorporated in a rule book. Employees operating this equipment must be trained and tested.

Safety Certification

LACTC will be self-certifying the safety of the LRT. For this purpose a self-monitoring safety auditing program is being developed by our consultant staff to verify that all practical steps have been taken at every stage of the design and construction to maximize operational safety. Not until all the required steps identified in the testing section are fulfilled will revenue service start.

IT'S SHOWTIME

It's July 4, 1990. The largest metropolis in the world without a mass transit system is about to lose that dubious distinction. Standing in back of the big ribbon is simply "a train" that represents the fruits of at least 70 years of planning, proposals, propositions, referenda, lobbying, debate, schemes, dreams, and a lot of hard work on the part of the people who pulled together to make it all happen.

If the Long Beach line is typical of light rail lines in general, it will carry about 1,000 percent of typical daily patronage that first day, partly because the price is right on that day only (free), but mostly because curiosity is riding at an all-time high.

The success or failure of this "show time" is a direct result of how carefully thought out the practice sessions were. How many rehearsals were done? To what degree was the testing carried out? Were the test data properly analyzed? Were deviations from the norm detected? Were corrective actions identified? Was retesting carried out to determine if the corrective action fixed the problem? If the trains prove unreliable, the reason will undoubtedly be traceable to one or more of the equipment suppliers. However, it will be LACTC, RTD, and the designer that will suffer the black eye. The final report card will be handed out by the media. If the image is favorable and the patrons perceive the rail system as reliable, then they will ride it and ridership statistics will be favorable as well.